THE SOCIAL CONSTRUCTION OF CLIMATE CHANGE

Global Environmental Governance

Series Editor: John J. Kirton
Munk Centre for International Studies, Trinity College, Canada

Global Environmental Governance addresses the new generation of twenty-first century environmental problems and the challenges they pose for management and governance at the local, national, and global levels. Centred on the relationships among environmental change, economic forces, and political governance, the series explores the role of international institutions and instruments, national and sub-federal governments, private sector firms, scientists, and civil society, and provides a comprehensive body of progressive analyses on one of the world's most contentious international issues.

Titles in the series

Governing Global Health
Challenge, Response, Innovation
Edited by
Andrew F. Cooper, John J. Kirton and Ted Schrecker
ISBN 978-0-7546-4873-4

Participation for Sustainability in Trade
Edited by
Sophie Thoyer and Benoît Martimort-Asso
ISBN 978-0-7546-4679-2

Bilateral Ecopolitics
Continuity and Change in Canadian-American Environmental Relations
Edited by
Philippe Le Prestre and Peter Stoett
ISBN 978-0-7546-4177-3

Governing Global Desertification
Linking Environmental Degradation, Poverty and Participation
Edited by
Pierre Marc Johnson, Karel Mayrand and Marc Paquin
ISBN 978-0-7546-4359-3

Sustainability, Civil Society and International Governance
Local, North American and Global Contributions
Edited by
John J. Kirton and Peter I. Hajnal
ISBN 978-0-7546-3884-1

The Social Construction of Climate Change
Power, Knowledge, Norms, Discourses

Edited by

MARY E. PETTENGER
Western Oregon University, USA

ASHGATE

Published by
Ashgate Publishing Limited
Wey Court East
Union Road
Farnham
Surrey GU9 7PT
England

Ashgate Publishing Company
Suite 420
101 Cherry Street
Burlington, VT 05401-4405
USA

Ashgate website: http://www.ashgate.com

British Library Cataloguing in Publication Data
The social construction of climate change. - (Global environmental governance)
1. Climatic changes - Environmental aspects 2. Climatic changes - Environmental aspects - Evaluation 3. Climatic changes - Government policy
I. Pettenger, Mary
363.7'3874

Library of Congress Cataloging-in-Publication Data
The social construction of climate change : power, knowledge, norms, discourses / edited by Mary E. Pettenger.
p. cm. -- (Global environmental governance)
Includes index.
ISBN 978-0-7546-4802-4
1. Climatic changes--Social aspects. 2. Climatic changes--Government policy. 3. Greenhouse gas mitigation--International cooperation. I. Pettenger, Mary E.

QC981.8.C5S626 2007
304.2'5--dc22

2006103145

ISBN: 978-0-7546-4802-4
Reprinted 2009

Printed and bound in Great Britain by
MPG Books Ltd, Bodmin, Cornwall.

Contents

List of Contributors

Karin Bäckstrand

Karin Bäckstrand is the Wallenberg Research Fellow in the Department of Political Science at Lund University, Sweden. Her research interests revolve around global environmental governance, the scope for transnational legitimacy and the role of scientific expertise in environmental policymaking. She received a Ph.D. from Lund University. Karin's research has been published in *Environmental Politics*, *Global Environmental Politics*, *European Environment* as well as in chapters in international book volumes.

Loren R. Cass

Loren R. Cass received his Ph.D. from Brandeis University, USA. He is an associate professor of political science at the College of the Holy Cross. His research focuses on the relationship between international and domestic environmental politics. He is the author of *The Failures of American and European Climate Policy: International Norms, Domestic Politics, and Unachievable Commitments* (2006).

Cathleen Fogel

Cathleen Fogel has led climate and other environmental policy work for a variety of non-profit organizations, including the Center for Resource Solutions, the Climate Group, National Wildlife Federation, Sierra Club and Greenpeace International. She holds a Ph.D. in Environmental Studies from the University of California Santa Cruz, USA, with research on biotic carbon sequestration policies under the Kyoto Protocol. She has published widely, including recent work in *International Environmental Agreements* (2005) and two reviews published by the Climate Group.

Takashi Hattori

Takashi Hattori is a Director (Program) for the Asia-Pacific Economic Cooperation (APEC) Secretariat. He received his Ph.D. from the Tokyo Institute of Technology, Japan. He is also a Consulting Fellow for the Research Institute of Economy, Trade and Industry, Japan. He has published numerous articles in *Interdisciplinary Environmental Review*, and in *International Negotiation*, as well as a chapter in *Innovations in International Environmental Negotiation.*

Myanna Lahsen

Myanna Lahsen received her Ph.D. in cultural anthropology from Rice University. She is a former Post-Doctoral Fellow with the US National Center for Atmospheric

Research and the Kennedy School of Government at Harvard University, where she also was Lecturer on Environmental Science and Public Policy. Presently she is Science Officer for Social Sciences for the Brazilian Regional Office of the International Geosphere-Biosphere Programme (IGBP) in Brazil, and Research Scientist with the Center for Science and Technology Policy Research at the University of Colorado at Boulder, USA. She has numerous publications, including book chapters, and articles in *Social Studies of Science*, and *Science, Technology, and Human Values*.

Eva Lövbrand

Eva Lövbrand holds a PhD in environmental science from Kalmar University, Sweden. She is currently a postdoctoral fellow in environmental politics at Lund University, the Swedish Centre for Climate Science and Policy Research (CSPR) at Linköping University, and the Center for Science and Technology Policy Research at University of Colorado. Her research on climate governance and the politics of scientific expertise has been published as journal articles in *Climatic Change*, *Global Environmental Politics*, and *Review of International Studies*.

Matthew Paterson

Matthew Paterson is Professor of Political Science at the University of Ottawa, Canada. His research focuses on the intersection of international political economy and global environmental politics. He has published widely in this field, including *Global Warming and Global Politics* (1996) and *Understanding Global Environmental Politics: Domination, Accumulation, and Resistance* (2000). He has recently completed a book called *Automobile Politics: Ecology and Cultural Political Economy* (2007).

Mary E. Pettenger

Mary E. Pettenger is an Assistant Professor of Political Science at Western Oregon University, USA. Her research interests include social constructivism, role theory and small states. She received her Ph.D. in International Studies from the University of Denver. She is currently researching active learning techniques in climate change negotiation role-playing simulations.

Heather A. Smith

Heather A. Smith is Associate Professor and Chair of the International Studies Program at the University of Northern British Columbia, Canada. She is also the Coordinator of Professional Development for the UNBC Centre for Teaching and Learning and a 2006 3M Teaching Fellow. Recent publications include a chapter (with Claire Turenne Sjolander) in *Canadian Foreign Policy* (Spring, 2005), in *Feminist Perspectives on Canadian Foreign Policy* (OUP, 2003), and in *A Decade of Human Security: Global Governance and New Multilateralisms* (Ashgate, 2006).

William D. Smith

William D. Smith is Assistant Professor of Anthropology at Western Oregon University, USA. He received his Ph.D. from Stanford University, USA. His recent field research has explored environmental history, transformations in peasant economies, and indigenous rights movements in the Sierra Norte region of the state of Puebla, Mexico. Currently, he is researching migration histories, colonial development, and urban ecology on the Mexico-US border.

Johannes Stripple

Johannes Stripple holds a licentiate of philosophy in environmental science from Kalmar University, Sweden, and a Ph.D. in political science from Lund University, Sweden. Johannes' research concerns the intersection of global environmental change and international relations, in particular questions regarding security, territory and authority. His work has been published in journals such as *Global Governance* and *Review of International Studies* as well as in several edited book volumes.

Foreword

People have been talking about climate change for a long time. They could hardly not do so once they realized that the glaciers had repeatedly advanced and receded on the earth's surface. If the earth had experienced a succession of warm and cool periods, then it seemed likely that today's relatively warm earth would cool and glaciers return some time in the future. Thirty years ago, some respected scientists suggested that cooling could take place suddenly and glaciation proceed quickly, and that this swing might be imminent. Stories in popular news magazines prompted a certain amount of public discussion of the coming ice age, but only briefly as more tangible and immediate concerns dominated the news.

The coming ice age never came. Yet no one claimed it was a 'social construction,' even if it was (and remains) a real possibility, however remote, and all the talk a real social phenomenon. A few years earlier, there had appeared a small book called *The Social Construction of Reality*, which its authors, Peter L. Berger and Thomas Luckmann, had given the daunting subtitle, *A Treatise in the Sociology of Knowledge*. Few scholars not specialists in an arcane branch of sociology were then familiar with this book; only gradually did it help to inspire a debate among scholars in many disciplines on whether, in what sense, and to what extent we make the world and its contents real. Ian Hacking assesses this debate in his aptly named book, *The Social Construction of What?* (1999), and it continues unabated. Still, speculative interest in a coming ice age, even today, would probably not inspire anyone to name a book *The Social Construction of the Coming Ice Age*, or even *The Social Construction of Climate Change*. Something else has happened in the last thirty years.

* * *

Indeed much has happened. Most obviously public discussion of climate change has returned with a vengeance. Large numbers of scientists from a number of disciplines are engaged in its study; news stories appear constantly; activists hold meetings; politicians make speeches and propose policies, which governments adopt, reject, implement or sabotage; governments negotiate with other governments. The scale and density of these connected activities have substantial effects on the world as we know it. This is exactly what large numbers of people intend by talking about climate change, and this must be social construction in any casual sense of the term.

People are not talking about global cooling, over which we would have no control (despite a flurry of concern in the 1980s over "nuclear winter" following a major nuclear war). They talk about global warming. A decline in global temperatures after 1945 had prompted the flurry of interest in climate change in the 1970s. When a century-long trend toward warmer temperatures resumed in the 1980s and then accelerated, even casual observers noticed a correlation with population, urbanization, industry, fossil-fueled electrification and transportation, and deforestation. Furthermore, science provided a plausible, highly generalized and comprehensible explanation for this correlation: the greenhouse effect. All sorts of activities associated with modernity release carbon dioxide into the atmosphere, which does not affect the amount of solar energy reaching the earth but does prevent radiant energy from returning to space.

Scientists know that the story is more complicated than this. Thus a warmer earth might increase cloud cover in turn increasing the amount of solar energy reflected away from the earth, with a commensurate cooling effect. Nevertheless, scientific opinion overwhelmingly favors the view that climate change in the last century is an unnatural event. We have done it to ourselves. We do not need scientists to tell us that we will have much to regret if the trend continues. We have done it to ourselves and we have only begun to pay. For many commentators, this is the Faustian bargain, modernity's deal with the devil, the inevitable consequence of our material excesses, our crimes against nature, our willful innocence, our invincible arrogance.

Modernity *is* a social construction (here again using the term in a casual sense). Climate change brought on by modernity tells us that social construction has material implications, often unintended in kind or scale. When our individual choices affect us collectively in unexpected and adverse ways, we call this result a problem. More specifically it is a collective action problem. Garrett Hardin's essay called "The Tragedy of the Commons" (1968) prompted popular awareness of this class of problems. The idea was hardly new; the title of the essay refers to overgrazing in pastures held in common.

Such resources are collective goods; nobody owns them. Nevertheless, unimpeded individual access may damage or destroy them. Even if we all know this, each of us still has a rational incentive to use the resource in question because we also know that everyone else has the same incentive. Anyone not using the good loses out and it ends up being devalued anyway. Some collective goods, such as clean air and an unchanging climate, we think of as natural. Some we construct and maintain: a pasture free for all to use as they see fit, collective personal security in a world Thomas Hobbes characterized as a war of all against all.

If the problem (depleted resources, collective insecurity) lies in the unintended consequences of our rational choices (to make and use things, be safe), then the obvious solution is an agreement to forego individual choice. Hobbes argued that society itself depends on such an agreement and the institutions needed to be sure that no one cheats (a rational choice that would encourage others to cheat; this is the collective action problem in a nutshell). Hobbes's social contract is, of course, a social construction: in this context the terms are all but interchangeable.

Collective action problems come in all sizes. Solving Hobbesian insecurity by creating states reproduces the problem in a world of insecure states. What to do about this problem is a question that absorbs an extraordinary amount of attention; the field of International Relations, to which this volume contributes most centrally, owes its existence to it. To say that climate change is a global problem is to plead for a big solution, a global social contract. The United Nations Framework Convention on Climate Change (1992) responds to this plea and, by doing so, defines the problem as one of a species: a big problem for states in a world of states. The Kyoto Protocol to the Framework Convention (1997) obligates states to reduce substantially "the emissions of greenhouse gases" within their territories. The agreement assumes that governments, as rational agents, can resist the temptation to cheat other states by not fulfilling their obligations. It also assumes that governments are able and willing to impose costs on their peoples to achieve collectively rational ends, when cheating (or not agreeing in the first place) is not only possible, but likely in the absence of the kind of institutions Hobbes had in mind for enforcing the social contract. In other words, defining global warming as a problem for states in the first instance means that states will play out their Hobbesian fears in yet another arena where failure could have tragic consequences. Greenhouse gases and nuclear weapons have unnerving conceptual similarities.

Is there an alternative in the world we have made for ourselves? A social contract is a top-down solution. The obvious alternative is the bottom-up solution that many social activists favor. Encourage public awareness that climate change is potentially catastrophic. Get people to change their habits, join with their neighbors, pressure their leaders. Large-scale social change can take place when people hear the right messages; states are part of the problem, not the solution. Recent history offers an inspiring example: the Wall came down. The problem for social activists is activating the right popular impulse at the right moment, and then controlling it.

Many of us are skeptical about top-down and bottom-up solutions to big problems. These problems are just too complicated not to require disaggregation and patchwork problem-solving. Yet this lengthy and uncertain process itself compounds the problem. We may not have time. Global climate change is already happening, and it may already be too late for any remedy. In the 1970s people began to worry about a global catastrophe due to overpopulation, resource exhaustion and irreversible environmental deterioration. A team of MIT computer scientists showed that exponential material growth has limits that may be exceeded even before growth ceases and anyone recognizes the need to act; unrestrained global growth will almost certainly lead to overshoot and collapse. Again, the idea is hardly new. Economists dismissed the MIT team as Malthusian pessimists who failed to realize that scarcity is an incentive for innovation and substitution: the market is the solution. Once the global recession of the 1970s passed, public anxieties eased and discussion subsided. Today the idea is back, captured in a ubiquitous simile: global warming is a runaway train, running on momentum, unstoppable, bound to crash.

* * *

That we can reduce the complex phenomenon of climate change to a few words and use a rhetorical flourish to give it enormous emotional power tells us that this phenomenon is a social construction and not just a collection of facts about temperatures, emissions, rainfall patterns and the like. Even the facts are for the most part social. Is it a fact that the earth's climate is changing, or is it merely fluctuating? Are we in fact responsible or not? Can we can do something or is too late? These questions can only be answered, if at all, by deciding first what we mean by such terms as "change" and "responsibility," by deciding who "we" are and what we might "do." When we say that something is a social construction, we are acknowledging that social facts are facts for social purposes, and that social reality is real even if it falls on us to make it real.

The contributors to this volume use the term "social construction" not just to indicate that large numbers of people think climate change is a monumental problem for humanity, or that we made this problem what it is, or that we have invested it with moral significance and emotional weight. They take social construction seriously enough to label themselves accordingly. They use the term "social construction" often but not at all casually because they have adopted a stance toward inquiry that we in the field of International Relations call constructivism. In other fields, we find scholars adopting a similar stance, sometimes calling themselves constructivists or social constructivists, sometimes social constructionists (as in the sociology of science). Yet every field is a different construction. In the field of International Relations, constructivists disagree on many issues, including who should be called or be calling themselves constructivists.

Constructivists agree that social facts are facts for social purposes, and that social reality is real. Not all of them agree with the converse claims that all facts are social facts because it is their social use that makes them facts at all, or that reality is only social. There are, "in fact," two kinds of constructivists in the field. As the editor, Mary Pettenger, points out in her Introduction, the two parts of the book differ in the philosophical assumptions of the contributors. She also says, quite correctly, that philosophical differences have political implications. The very fact that the book is organized in two parts tells a story about constructivists and how they see themselves divided.

In Part I, contributors influenced by functionalist sociology focus their attention on social construction as a normative phenomenon. What we want as social beings affects what we do, and, it must be added, what we want inevitably includes the way we think things should be. Norms or rules convey what we think people should do and give society the wherewithal to coordinate what we do individually and to fulfill our wants and needs. The contributors to Part I investigate processes by which norms emerge and change (often as an unintended consequence of what we do collectively), empower agents and form institutions, and in the process change the world. Climate change is obviously normatively charged as a social construction, and focusing on norms constitutes our best chance to change the trajectory of climate change.

Contributors to Part II are no less interested in the normative properties of social construction. They are less interested in norms as things, however, and more inclined to view social construction as a process inextricably associated with all the talk we engage in. Not only does this talk produce norms affecting what we do (and talking is doing), but talking is always to some end. We want what we say to have an effect on others. Influenced by poststructuralist literary and social theory, Part II's contributors focus attention on the persuasive properties of speech and on discourse as the power-suffused result on many people speaking to each other. What we do about climate change obviously depends on the stories we tell. As these stories change, the world changes too.

Social construction is always political, whether people think so or not. Despite their differences, constructivists agree among themselves that social facts are real because we treat them as real. This "fact," as constructivists see it, translates into an emphasis on processes. The many scholars in the field who are not constructivists tend not to attach the same significance to processes and their general properties. This is a telling difference with political implications. For constructivists, social processes always have a material dimension and constitutive effects. That these processes are social links them to other such processes. Through these many, linked processes we, as social beings, give meaning to the world. In turn the world made meaningful makes us the social beings that we are. As a social construction, climate change is no one thing. Instead it is an ensemble of constitutive processes, yielding an ever changing panoply of agents and institutions, fixed in place and meaning only for the moment.

Nicholas Onuf

Acknowledgements

This project was conceived at an informal meeting at the International Studies Association convention in Portland, Oregon in 2003. From then to now, the group has been graced by the presence of many people who have influenced its evolution, some for only a short time, and others for the duration. We say a special thanks to those that joined the project, provided us their insight and support, but had to leave the project due to unforeseen circumstances. I personally thank all of the contributors for their hard work, patience, and willingness to join such a challenging project. Special thanks to Loren Cass for his support and perseverance since the very beginning of the project. The journey has been a reward.

I am grateful to Western Oregon University for providing funding for this project, and to Stephanie Hampton and Becky Myrold who served as my research assistants. Thank you to John Kirton, the series editor for the Ashgate series on *Global Environmental Governance*, his assistant Madeline Koch, and to the four anonymous reviewers for their valuable insights that have served to improve the book. We would like to extend thanks to *Global Environmental Politics* for allowing Karin Bäckstrand and Eva Lövbrand to reprint portions of their article. In addition, the advice and comments from Jack Donnelly and Nick Onuf have been indispensable, and I owe them my sincere gratitude. And finally, I speak for all the contributors in giving a special thanks to our families and friends who have provided us with support throughout the project and remembrances for those that are no longer with us. We hope that you enjoy reading the book and find it valuable in many more ways than we can ever anticipate.

List of Abbreviations

AAU	Assigned Amount Units
ACBE	Advisory Committee on Business and the Environment
AGBM	Ad Hoc Group on the Berlin Mandate
AIM	Asia-Pacific Integrated Model
ANT	Actor Network Theory
AOSIS	Alliance of Small Island States
BDI	The Federal Association of German Industry
BuZa	Ministry of Foreign Affairs (the Netherlands)
C & C	Contraction and Convergence
CAFÉ	Corporate Fuel Economy
CalPERS	California Public Employees Retirement System
CalSTRS	California State Teachers' Retirement Systems
CAN	Climate Action Network
CARB	California Air Resources Board
CASA	Citizens Alliance for Saving the Atmosphere and the Earth
CCP	Cities for Climate Protection
CCX	Chicago Climate Exchange
CDA	Christian Democratic Appeal Party (the Netherlands)
CDM	Clean Development Mechanism
CDP	Carbon Disclosure Project
CDU	Christian Democratic Union (Germany)
CFCs	Chlorofluorocarbons
CH_4	Methane
CIESIN	Consortium for International Earth Science Information Network
CO_2	Carbon Dioxide
COP	Conference of the Parties
CSE	Center for Science and Environment
CSU	Christian Social Union (Germany)
ECP	Eastern Canada Premiers
EEA	European Environment Agency
EEN	Evangelical Environmental Network
EPA	Environmental Protection Agency (United States)
ETS	Emission Trading Scheme
EU	European Union
G77	Group of 77
G7/8	Group of 7/8

GEF	Global Environmental Facility
GEP	Global Environmental Politics
GHG	Greenhouse Gases
HFC	Hydrofluorocarbons
IAI	Inter-American Institute
ICC	Inuit Circumpolar Conference
ICCI	Illinois Conservation and Climate Initiative
ICCN	Interfaith Climate Change Network
ICLEI	International Council on Local Environmental Initiatives
ICSU	International Council of Scientific Unions
IET	International Emissions Trading
IISD	International Institute for Sustainable Development
INC	International Negotiating Committee
INCR	Investors Network on Climate Risk
IPCC	Intergovernmental Panel on Climate Change
JI	Joint Implementation
LCPD	Large Combustion Plant Directive
LDC	Less Developed Country
LDP	Liberal Democratic Party (Japan)
LPF	List Pim Fortuyn Party (the Netherlands)
LULUCF	Special Report on Land Use, Land Use Change and Forestry
METI	Ministry of Economic Trade and Industry (Japan)
MEZ	Ministry of Economic Affairs (the Netherlands)
MIT	Massachusetts Institute of Technology
MITI	Ministry of International Trade and Industry (Japan)
MoE	Ministry of Environment (Japan)
MoFA	Ministry of Foreign Affairs (Japan)
MOP	Meeting of the Parties
MW	Megawatts
N_2O	Nitrous Oxide
NAFTA	North American Free Trade Agreement
NASA	National Aeronautics and Space Agency
NEAA	The Netherlands Environmental Assessment Agency
NEG	New England Governors
NEPP	National Environmental Policy Plan
NO_x	Nitrogen Oxides
NRC	National Research Council
NRDC	National Resources Defense Council
NRPE	National Religious Partnership for the Environment
OECD	Organization for Co-operation and Economic Development
PFC	Perfluorocarbons
PG&E	Portland General and Electric
PHEV	Plug-in Hybrid Electric Vehicles
PvdA	Labor Party (the Netherlands)

RAN	Rainforest Action Network
RECs	Renewable Energy Credits
RGGI	Regional Greenhouse Gas Initiative
RGGR	Regional Greenhouse Gas Registry
RIVM	National Institute of Public Health and Environmental Protection (the Netherlands)
RPS	Renewables Portfolio Standards
SAF	Set America Free
SAR	IPCC - Second Assessment Report
SF_6	Sulfur Hexafluoride
SMIC	Study of Man's Impact on Climate
SO_2	Sulfur Dioxide
SPD	Social Democratic Party (Germany)
SUVs	Sports Utility Vehicles
UN	United Nations
UNCED	United Nations Conference on Environment and Development
UNEP	United Nations Environment Programme
UNFCCC	United Nations Framework Convention on Climate Change
USCCB	United States Conference of Catholic Bishops
VOCs	Volatile Organic Compounds
VROM	Ministry of Housing, Spatial Planning, and Environment (the Netherlands)
VVD	Liberal Party (the Netherlands)
WCGGWI	West Coast Governors' Global Warming Initiative
WGA	Western Governors' Association
WMO	World Meteorological Organization
WREGIS	Western Renewable Energy Generation Information System
WRI	World Resources Institute

To our families and friends

Chapter 1

Introduction: Power, Knowledge and the Social Construction of Climate Change

Mary E. Pettenger[1]

> Kyoto is without doubt only the first step. We will have to do more to fight this rapid increase in temperature on our wonderful blue planet earth.
>
> Klaus Toepfer, Head of UNEP, 2005

> No problem can be solved from the same level of consciousness that created it.
>
> Albert Einstein

Climate change has come to have numerous meanings "on our wonderful blue planet earth" (Toepfer 2005). Over the last twenty years, small steps have been taken to recognize, conceptualize and address climate change while scientific, political, economic and rhetorical battles rage. Recently we have taken tangible actions concerning climate change as illustrated in February 2005 when the Kyoto Protocol came into force. Yet, the treaty, requiring a binding commitment by member states to change their climate impacting behavior, has faced enormous criticism from multiple sources during its construction, ratification and implementation phases. The existence of climate change, its potential impacts and even the efficacy of current measures remain in contention.

Peter Haas (2004) presents an interesting and important question: "When does power listen to truth?" In his essay he discusses the means by which truth, as generated by science, influences political processes. Haas examines the social processes responsible for generating "usable knowledge" with which new realities can be birthed regarding climate change. Thus, he is addressing the age-old dialogue between power and knowledge. When do power and knowledge subvert or unite with each other? And, when does knowledge achieve its own power and bring change? In essence, the *Social Construction of Climate Change* asks these same questions. We are seeking to expose the dialogue between power and knowledge found in the social construction of climate change.

Taking on such a lofty endeavor has its obvious problems which we have actively sought to identify and address with varying degrees of success, but no book can be everything for everyone. When we first began to discuss the focus of the book,

1 I wish to thank Nicolas Onuf, Jack Donnelly and Peter Callero for their comments and assistance on the introduction, and for their very detailed and helpful suggestions.

the contributors agreed that the policies to address climate change to date were problematic, and we consciously chose to cast a broad net to uncover vectors of change that might bring new understandings of climate change politics. The chapters in this book do not focus explicitly on the questions of what constitutes climate change. Nor do they explicitly examine how, or if, we should respond. Instead, we are exploring the construction of the questions. More precisely, what does it *mean to ask* if climate change exists? How has the process been socially constructed that interprets the existence of and responses to climate change?

Most of the contributors to this volume are scholars in the field of International Relations (IR). The field, while varying individually in theoretical orientations and assumptions, focuses on the organization, use and effects of power in its many forms. More recently, some in the field have begun to focus on the role of knowledge, and those producing this knowledge, as a form of power. For example, Peter Haas examines the role epistemic communities may play in generating knowledge, gaining power and influencing political processes (Haas 2004). In this book, we trace the dialogue between power and knowledge within which the social understandings of climate change have been constructed. Through our discussions and in our writings, we seek greater knowledge about climate change, but not always in a problem-solving manner that is the norm in IR. Problem-solvers generally assume that factual statements about the world can be tested and proven to be true or false based on principles found in natural science. This approach (often called positivism) assumes that facts exist outside of the observer who remains impartial in the process, and that all results can be subject to testing, leading to a systematic and reliable progression of knowledge.

In contrast, the contributors to this book in various ways and degrees challenge the assumptions of positivism and the categorization of climate change as a "problem" that can be analyzed and solved. To this end we have adopted the IR approach of constructivism. While the contributors support different perspectives on constructivism within the book, we all share much in common. We consciously choose to focus on several regions of the world, and to include northern/western and southern/indigenous voices. We are seeking to understand the interpretations of climate change, vis-à-vis the social processes of climate change conceptualization, i.e., what *it is* and how its causes and consequences, and the planned responses to it, are constructed.[2] Additionally, we specifically are exploring change. The book focuses on the processes of social construction of climate change, primarily from the late 1980s when climate change became an increasingly salient issue to the present.

2 This introduction foregoes the role of a detailed primer on the mechanics, causes and consequences of climate change, and much detail is included in the chapters themselves. In addition, a large body of literature explores climate change, including natural sciences/ scientific studies and in several social science disciplines. In the field of international relations, climate change has been heavily studied (see for example, Adger et al 2006; Dessler and Parson 2006; Cass 2006; Fisher 2004; Faure et al 2003; Luterbacher and Sprinz 2001; Miller and Edwards 2001; Newell 2000; Paterson 1996; Rowlands 1995, as well as an enormous amount of articles and chapters).

It is through this process of change that we hope to identify the emerging forces arising from the structures of power and knowledge.

While some might judge the book for doing too much on a cursory level, our intent is to look broadly in order to uncover the hidden meanings lost in the climate change arena. At the same time, the cases and themes in the book touch on other areas in the global environmental politics (GEP) literature. What makes this book different is that we adopt the lens of constructivism to view the cases, and place the cases together within this theme to seek information that is not found through other theoretical and empirical processes. While the chapters are comparative in nature, we are not seeking to compare policy formation in different states and arenas.[3] Rather, we have two main goals, to illuminate new understandings of climate change, and to compare two different constructivist perspectives. In the end, we are seeking new questions to ask that might gently prod future climate change negotiations and policies. We suggest that new "levels of consciousness" are required to address the substantial "problem" of climate change. What follows serves to guide the reader in understanding the purpose, theoretical roots and structure of the book.

Climate Change Meanings

The definitive assumption of this book is that climate change must be understood from the context of social settings; however, the importance of how (and for some, why) knowledge and power structures are generated is often ignored by policymakers and academics. The following discussion of the role of science in climate change policy illustrates this obfuscation. Scientists are bound within the realm of the scientific method to seek causality between human actions and climate change. However, even though there is an overwhelming amount of empirical evidence to support the existence of climate change, uncertainty remains (for an important discussion of this topic, see Oreskes 2004).

In response, there is a growing body of literature investigating the interactive role of science and the social construction of knowledge (Lahsen this volume; see also Lahsen 2005; Bäckstrand 2004, Demeritt 2001; Rosa and Dietz 1998). The construction of knowledge is fundamental to understanding climate change, and yet overwhelming evidence/knowledge has not led to power as Haas (2004) presents. The controversy continues when policymakers such as Schlesinger (former US Secretary of Energy) state the following about climate change science:

---

3 There is a considerable amount of valuable material in the comparative environmental policy literature, and more specifically literature that addresses climate change. We direct interested readers, in addition to the numerous sources cited in the Introduction and the chapters in the book, to the following: (1) other volumes in the Ashgate Global Environmental Governance Series, and (2) comparative literature (Spaargaren et al 2006; Bauer 2006; Jasanoff and Martello 2004; Conca and Dabelko 2001; Litfin 1998; Hajer 1995).

> ... science is not a matter of consensus, as the histories of Galileo, Copernicus, Pasteur, Einstein and others will attest. Science depends not on speculation but on conclusions verified through experiment. Verification is more than computer simulations—whose conclusions mirror the assumptions built in the model. Irrespective of the repeated assertions regarding a "scientific consensus," there is neither a consensus nor is consensus science (Schlesinger 2005, A10).

Consequently, the perceived material reality of climate change is defined in social settings by scientists and policymakers (who may or may not be experts, Lahsen 2005). In other words "science ... *is* the politics of climate change" (Lahsen this volume, p. 190).

Concurrently, while scientists work to make sense of climate change, some world leaders seize upon this brief gap of uncertainty to obstruct efforts to mitigate climate change. Are these political leaders uneducated or naive, or is there something deeper taking place? How can we understand the myriad of responses to climate change? For example, for what reason(s) has the largest contributor of greenhouse gases, the United States, chosen to impede international cooperation? How has policy emerged that has negated the urgency for action and/or obscured this issue in relation to other issues? These and other such illustrations and questions flow throughout this book.

At the same time, what role do other powerful actors, such as the media, play in our understandings of climate change? For example, the US infotainment industry increasingly has portrayed divergent climate change interpretations. A recent Hollywood movie, *The Day After Tomorrow*, sensationalized a cataclysmic, climate change event (Emmerich 2004). The film stirred controversy by portraying the climate of the world altering in just a matter of days, with much of the Northern Hemisphere covered in ice because of global warming. In contrast, *State of Fear* by Michael Crichton presents climate change as a hoax perpetrated by those who would use misperceptions of science to gain political and economic power. Crichton's portrayal of global warming based upon his own scientific research (see his chapter "Author's Message") weaves an impressive amalgamation of science and entertainment (Crichton 2004).

Additionally, the documentary about former US Vice President Al Gore and his climate change awareness speeches, *An Inconvenient Truth*, presents worst-case-scenarios, including a twenty-foot sea level rise, and is an unbridled call for action (Guggenheim 2006). The movie has generated a large following of devotees in the US who flock to hear his tale and push for political changes amidst counter-attacks by opponents who claim he is distorting reality and seeking the political spotlight. What is climate change and how have our understandings of it been socially constructed? The general public is bombarded with opposing interpretations of climate change by the media, while facing increasingly more threatening physical/material realities.

In recent years it seems that the number of severe weather disasters has increased. Drought in South Asia, floods in Europe, typhoons in East Asia, and hurricanes in the Atlantic Ocean left death and suffering in their wake. While these phenomena are not new, the severity of their occurrences has changed. The number of wildfires are on the rise and their increase has been linked to "higher temperatures" (Hotz 2006).

Glaciers are melting at an unprecedented rate all over the earth (Blue Earth Alliance 2006). 2005 was the hottest year on record for the last one hundred years, with 1998, 2002, 2003 and 2004 following on the list of the five hottest years (NASA 2006). Are all these material facts interrelated? How should climate change be interpreted and understood? Should we interpret these facts, as did Al Gore in *The Inconvenient Truth*, to convey the potential annihilation of the human species if we do not act immediately? Or should we respond as President Bush of the US has done since 2001, proclaiming that the scientific evidence is too uncertain and the economic costs are too great to require an immediate response?

Ask ten people how to define climate change, its causes and effects, and you will get ten different answers. The language used to discuss and describe climate change is often value-laden as the terms employed have different meanings depending on who is discussing the topic and why. What is clear is that the meaning of climate change is defined in social settings. The range of terms used to conceptualize the perceived material reality (of the rising mean surface temperature of the earth caused by increasing greenhouse gases) extends from global warming and the greenhouse effect, to others who might add a twist such as climate hoax, climate crisis and climate dilemma.

This book adopts climate change as a conventional term for several reasons. First, climate change is the term that was selected during United Nations negotiation for the United Nations Framework Convention on Climate Change (UNFCCC). Second, the choice of climate change implies certain subjective meanings. Global warming for some is a more explicit description of the process that is perceived to emerge out of the anthropogenic build up of greenhouse gases in the atmosphere and the resulting rise in mean surface temperature. Yet, this term often creates a misconception that global warming will result in hotter weather all over the world when changes in climate will take varying degrees and forms all over the world. We have selected climate change for conventional as well as practical purpose, while recognizing that the concept carries its own pejorative baggage.

Now look deeper into this picture at the different political agendas of the participants. Each has diverse answers for the following questions. Does climate change exist? Who or what is causing it? Can we predict the results of climate change? What actions can best reduce climate change? Who should pay for such changes? Is climate change a myth, hoax or, perhaps are those who claim the existence of anthropogenic climate change proposing a new "theology" as recently suggested by Schlesinger (2005, A10)? These types of questions are perplexing and yet interesting. Clearly, definitions and debates on climate change dance across the realms of science and politics. We are interested in exploring these questions from the social bedrock upon which understandings of climate change are formed, thus we have selected constructivism as our framework. I now turn to a brief overview of constructivism, the two theoretical perspectives adopted in the book, and the controversy surrounding their inclusion within constructivism.

Climate Change and Constructivism

Social scientists increasingly have adopted constructivism to understand environmental issues in general (see for example Broadhead 2002; Hannigan 1995) and more specifically, as a framework to examine the ocean (Steinberg 2001), nature (Eder 1996; Hajer 1995), and climate change (Cass 2006; Oels 2005; Hoffmann 2005; Demeritt 2001). We seek to build on this growing and useful body of literature.

Power, Knowledge and Constructivism

Conceptually, the book is organized by three interlinked constructivist principles nested within the broad theme of power and knowledge. 1) Privileging material *and* ideational forces allows for, (2) giving primacy to agents *and* structures, and thus (3) opens the view to expose processes and change. What follows is a brief discussion of these three principles with the note that dividing them into three is artificial.

Ideational/Material Factors The first principle of constructivism "denies ontological primacy either to ideas or social categories (e.g., war and peace), or to material things (guns, butter). Instead, we hold that social facts (such as war) are real because they always have material consequences, and that material things (such as guns) are real by virtue of social construction" (Onuf 2006). Constructivists, with intent, argue that "the material and ideational are complexly interwoven and interdependent" (Hay 2001, 7). As such, any study of climate change must give value to both. This approach does not negate the power of material realities, but rather, assists in the understanding of how material realities gain meaning through social interaction. For example, as discussed in the beginning of this chapter, interpretations of climate change are shaped by social and physical/material forces. Or, as William Smith notes in his chapter, there would be no climate change (as we know it) without the main greenhouse gas, carbon dioxide (CO_2). Each chapter touches on this theme to various degrees and from different perspectives. For example, while Paterson and Stripple study the social definitions of climate change as a threat or an opportunity, Lahsen examines distrust by Southern scientists of the social construction of climate change knowledge.

Agent/Structure Duality The second principle of constructivism this book highlights is the agent/structure duality. Constructivists (see for example, Hopf 2002; Ruggie 1998; Kubálková et al. 1998; Finnemore 1996; Katzenstein 1996; Onuf 1989; Kratochwil 1989; Wendt 1999), with varying degrees and points of emphasis, focus on agent and structure, and search for understanding through underlying intersubjective meanings embodied in identities, interests, and structures such as norms and discourse. In allowing ideational forces, actors gain agency (the ability to make choices as social beings interacting within a structure) and the agent/structure duality becomes recursively co-constituted. For example, constructivists contend that the behavior of states is not always determined solely by their materially-defined power; they can construct reality based on socially-defined, intersubjective meanings (Wendt 1992). In addition, the social context generated from human consciousness and interaction is a structure that constitutes, as well as restricts and enables, human actions (Adler 1997, 327).

The social construction of actors identities and interests, and of structures such as discourses and norms is the heart of constructivism. Thus, this approach is germane to the book as the interests and identities of climate change actors (individuals, substate actors, states, and international organizations) emerging in the UNFCCC Conference of the Parties (COP) negotiations, domestic politics and state interactions are fodder for bringing greater understanding of climate change policies. This issue is integral to understanding the social construction of climate change, because vis-à-vis the processes of knowledge construction and power struggles, some actors are privileged while others are negated. As such, we are asking what structural forces are at work, which agents have power, and what role does knowledge play in this social interaction?

Process and Change Finally, in giving primacy to the material and ideational, and agents/structures, constructivism holds promise to understand change. We consider this end to be one of the most valuable contributions of this book. Through social processes (not always determined by material realities but sometimes constructed in social settings by actors, and constrained and enabled by structures such as discourses and norms), change is made transparent. As noted by Burch: "Constructivism shifts attentions away from objects (actors, structures) to processes (constitution, construction, creation, learning) ..." (2002, 62). Consequently, constructivism and climate change appear well suited for each other. Actions regarding climate change have appeared and fluctuated significantly for over twenty years, thus providing rich and available material. At the same time, the process of policy formation has led to less than desirable results because of a wide variation in the interests and identities of the actors involved.

Concomitantly, constructivism allows us to view climate change from a new perspective with the hope of uncovering processes, actors and structures that have been obscured in the current framing of climate change. In this book, the concept of usable knowledge (Haas 2004) is central to understanding the process of international negotiations, and the disparities between knowledge and power. We are led to the question, what is it that we truly know about climate change, and how have we come to know this? How has the policy process developed, including the recognition that such change is not necessarily linear (as in progressive or regressive)? What has changed and why? In short, constructivism offers important guidance with which to answer the questions of this study because it recognizes ideational as well as material forces. In addition, it "emphasize[s] the construction of social structures by agents as well as the way in which those structures, in turn, influence and reconstruct agents" (Finnemore 1996, 24). In other words, it leaves space to expose processes of change. The chapter now turns to discussing the two perspectives of constructivism adopted in this book.

Norm-centered and Discourse Analytical Perspectives

The book consciously does not assume one constructivist approach but rather two, a norm-centered and a discourse analytical perspective. Each represents a different constructivist approach and offers similar, and yet importantly divergent portrayals of climate change. Thus the book is intentionally seeking not only individual interpretations of the social construction of climate change, but also different interpretations of constructivism itself. Additionally, we are seeking a dialogue between the norm-centered and discourse analytical perspectives within constructivism. We also sustain the advice of Zehfuss: "We must recognise our decision for what it is: a decision. Anything else would imply irresponsibility passing as responsibility" (Zehfuss 2002, 249), and that "treating it [our decisions] as such is political" (ibid., 246), but "we must create a beginning" (ibid., 248). Our intent is to include a broad range of constructivist text to expose the processes of social construction of climate change and not to exclude approaches based on our own subjective desires and predilections. One important effort made by the contributors to this book was to read and comment on the other chapters of this book. In essence, we too are engaged in the social construction of climate change and as such have pushed ourselves to view the issue from new and sometimes contradictory perspectives.

A review of the constructivist literature in IR quickly yields the perception that one cannot write about the topic without raising the critical issues of the philosophical assumptions embedded in the study. The assumptions significantly flavor the methodology employed and outcomes produced (Hay 2001), and each choice made implies political interests as well as a philosophical stance. Clearly, constructivists take debates over positivism and its alternatives seriously (for a useful summary, see Green 2002). It is not the intent here to reconstruct all these debates and their implications for constructivism, but rather to illuminate some important tensions within the field and within the book. Consequently, we recognize that some readers will disagree that one part of the book is an adequate representation of the social construction of climate change, while others will reject the other part. We recognize, in addition, that all representations including the ones in the book, are politically charged social constructions.

Clearly what constructivism means is interpreted differently within the field itself. Several articles have dealt with these differences, presenting two (Fierke and Jørgensen 2001; Christiansen et al 1999), three (Reus-Smit 2005; Zehfuss 2001; Burch 2002), or perhaps even four (Adler 2002) different constructivist approaches. For example, Burch identifies three "versions of constructivism": "structure-oriented," "norm-oriented" and "rule-oriented" (2002, 64–69). In contrast, Zehfuss distinguishes between Wendt, Onuf and Kratochwil's "constructivisms" and outlines the principal role that language plays for both Onuf and Kratochwil in contrast to Wendt (Zehfuss 2001).[4]

---

4 Those conversant in the different perspectives or versions of social constructivism might wonder why we adopted only two of the perspectives present in the current literature

An article in *International Studies Review* directly illustrates this divide. The article encompasses a debate between Checkel and Dunn, with mediation by Klotz and Lynch. In essence, Klotz is seeking to bridge the divide between norms and discourse approaches in constructivism. Checkel presents a rational-positivist constructivism and focuses on causality, while Dunn favors an interpretive constructivism and exploring the production of meanings. Klotz and Lynch attempt to uncover common language and approaches. In the end, there are some strong commonalities (thus making what constructivist approaches share), but there are some strong differences that might make crossing the divide insurmountable (Klotz and Lynch 2006; Checkel 2006; Dunn 2006).

Our volume adopts the main precept of constructivism, constructivists in the field of international relations focus on social forces to explain or understand the process(es) of political change. And, to gloss over a great deal of substance in the field, we recognize that most constructivists attempt to find a "middle ground" between empiricism (positivism) and pure subjectivity (Fierke and Jørgensen 2001; Adler 1997).[5] In the spirit called for by Inayatullah and Blaney (2004), we are seeking the other by critically allowing differences and embracing the divide. What follows is a brief overview of the two constructivist approaches found in this book and the dilemmas in attempting to apply both within the same setting.

Norm-Centered Constructivism

Norm-centered constructivists focus on norms as set expectations to which agents attach some sense of obligation for themselves and others (Checkel 2005; Björkdahl 2002; Farrell 2002; Payne 2001; Cortell and Davis 2000; Katzenstein 1996; Finnemore 1996).[6] The norm-centered approach has been referred to as soft constructivism as it leans more heavily toward rationalism and a positivist methodology. Norms are employed as conceptual tools to explore the social construction of international politics, e.g., "states are guided by norms that define the identities of the main actors in world politics ... and define the formal rules and accepted practices of the international game" (Farrell 2002, 52). As such, norms

(including perhaps rule-based perspectives). The absence of these alternatives was not done purposively as the original intent was to present additional perspectives. However, due to the vagaries of edited volumes, the chapters were created by scholars who divided themselves into norm-centered and discourse analytical perspectives.

5 The introduction, for the sake of brevity, does not address the issue within the social constructivist field of whether social constructivism is a theory, metatheory or analytical framework. For the purposes of this study we are assuming that social constructivism has an ontological basis by privileging the social and the material, while it also acts epistemologically as an analytical framework, while realizing that this project may conflate and avoid the issue. In addition, as an edited volume, we are focusing more directly on the social construction of climate change, rather than on theory building.

6 Readers are encouraged as well to review Chapter 2 by Loren Cass who provides an excellent overview of norm-centered constructivism.

appear as structural phenomena, guiding meanings that exist as the other, external to the actor, and yet are recursively defined by actors in a social setting. Thus, norms are a social construct, created by actors interactions, arising with intersubjective meanings and, recursively coming to constrain, empower and define actors' social understandings.

Even if norms are not obeyed, they determine the proper, "ought to" behavior of agents based on agreed on, or legitimated, sets of behavior (Ruggie 1998, 97). Norms have, in essence, a causal force that can shape actor's behavior in that they socialize actors to the proper behavior. However, one of the major critiques of norm-centered constructivism is that it lacks an adequate explanation of agency, i.e., how norms are transformed to action, and how and why actors choose to follow or reject certain norms. Recent efforts in the norm-centered approach (see Finnemore and Sikkink 2001; Checkel 1997) have sought to place greater emphasis on agency as do the chapters in this volume.[7] In keeping with positivist assumptions, constructivists who adopt a norm-centered perspective assume causal forces present in the processes of social construction. Thus, these studies seek to explain how and why actors define, adopt and instantiate particular norms. Part I of this volume illustrates this perspective in examining the processes of norm creation, adoption and diffusion.

Discourse Analytical Perspective

The discourse analytical perspective is based on the same tenets of constructivism but is radically different in its assumptions, intents and methodology. This perspective leans more heavily toward critical theory and a rejection of the positivist methodology, as it seeks to uncover how shared meanings are privileged or marginalized in social settings (Oels 2005; Debrix 2003; Zehfuss 2002; Fierke 2001; Paterson 2001; Milliken 2001 and 1999; Weldes 1998; Hajer 1995; Litfin 1994; Doty 1993; Onuf 1989). In other words, identifying and examining the causal forces present to determine *why* actors behave in certain manners would be antithetical to most discourse scholars. In contrast, those who adopt a discourse approach seek to interpret *how* discourses emerge, and uncover the process of power and knowledge formation. They emphasize the complexity of social reality and the impossibility of isolating proximate causes in the continuous co-constitution of agents and their normative practices.

Thus, this approach seeks to lay bare the dynamics of power and privileging of actors, examines the role of language and, more importantly, the processes surrounding the use of language, "to include symbolic systems, institutional structures, and social rules and practices" (Livesey 2002, 122). Discourse is more than simply the use of language as a tool for communication. Discourse conveys subjectivity, knowledge and power, and "... discursive practices situate actors (including both individuals and

7 We recognize as well the difficulties of giving primacy to both agent and structure. Uniting agent and structure may be "the middle [that] represents a black hole, into which social theories and philosophies vanish without trace" (Hollis 1994, 257).

organizations) in matrices of power, which privilege some interests and marginalize others" (ibid., 123). As such, the discourse analytical chapters seek to illuminate the deeper, surrounding processes within which the social construction of climate change takes place.

In sum, the book focuses on exposing dominate norms and discourses, as well as uncovering marginalized/alternative norms and discourses present in the climate change arena. These chapters provide an important overview of the interaction of power (who or what has dominated the process) and knowledge (what information is privileged and negated by those who have power). In addition, they offer important portrayals of marginalized/alternative interpretations of climate change that may or may not rise to prominence in the changing processes of the social construction of climate change. In essence, we are postulating that new norms and discourses may emerge from latent voices.

Constructivism enables us to uncover how international structures and actors, and their subsequent actions, remain constant and change. It illuminates what actions are acceptable and unacceptable for international actors. Most importantly, it can lead us to understand how certain meanings have emerged and been framed, while others have been obscured. The theme of power and knowledge which contains these three principles of constructivism will be returned to in the Conclusion, while the reader is urged to keep these three principles in mind while reading the chapters. Finally, this chapter turns to an overview of the book.

Recurring Themes and Overview

All of the chapters are united around exploring and presenting different pictures of and voices within the social construction of climate change. The chapters as a whole focus on different levels/scales and cases, and yet they share two important commonalities. First, it is clear that powerful actors operating at the local, state, and international levels have profoundly shaped the process of climate change politics. For example, developed states have dominated international climate politics and within many countries economic actors have imposed their will on climate policy.[8] However, the processes by which this domination has taken place gains greater transparency when the transfigurations of power and knowledge are uncovered, as each of the chapters evaluates the relationships among power, material forces, and ideas in the social construction of climate change.

Second, based on the chapters, it appears that many forms of socially constructed climate change knowledge are currently present. While there is a constant struggle between norms and between discourses (and between norms *and* discourses), this struggle does not appear to be either linear or determined. In other words, new voices

---

8 We would like to note that we did not set out from the beginnings of the project to include only developed states as cases in the norm-centered section. However, this focus has yield some interesting results about climate change policy and comparisons between the two parts of the book as will be discussed below.

are constantly emerging in the processes. And if we assume, as constructivism does, that change is possible (and potentially inevitable), then these alternative voices may become more prominent, or perhaps absorbed into the prevailing norms, discourses and policies. Based upon these two points, the following section presents two themes.

The first theme relates to the processes of change in the social construction of climate change. One of the most important contributions of constructivism is the ability to illustrate changes in values, identities, interests, strategies and policies that a focus on material forces alone cannot explain. One important question to be explored is how (and why) dominant norms and discourse gain prominence. How do they gain power and configure knowledge? Who or what enables this process? In addition, how do marginalized norms and discourse challenge the dominant forces, and perhaps gain saliency? For example, why and how have the ecological modernization discourse and the economic efficiency norm dominated climate policies? Is it possible that the reform civic environmentalism discourse and the stewardship norm can rise to replace liberal economic discourses and norms? Answers to these questions can be found by exploring the social construction of climate change, in addition to requiring an analysis of norms/discourses and agency. In this book, we begin to identify who the critical actors are and how we might explain or interpret their ability to shape the dominant norms and discourses.

A second set of themes relates to several sources of change presented in the chapters. These sources may serve as spaces within which to initiate challenges to the dominant discourses/norms. Four spaces emerge from the chapters, briefly mentioned here and further discussed in the Conclusion. First, the chapters explore how the *interaction between nature and humans* affects our understanding of the dominant norms/discourses. Second, they ask what role *the construction of knowledge through science* plays for climate change. Third, they discuss *the role humans as social* (rather than merely political or economic) *beings* have on the future of climate change policy. Fourth, they delineate the theme of *participation* that resonates throughout most of the chapters in the book, asking how participation affects the process through which norms/discourses emerge and achieve influence. I encourage the reader to keep these themes and topics in mind while reading the book.

Finally, as this is only a small step in adopting constructivism as a framework to explore climate change, the book cannot contain all possible perspectives. For example, some might prefer a larger norm or discourse section. Others might critique the book for lacking cases such as an African state, or China or India that will emerge soon as major CO_2 emitters. We urge others to learn from our efforts and to generate further studies, comparisons and constructive suggestions. A short synopsis of each chapter follows.

Part I is comprised of norm-oriented chapters. Chapter 2, by Loren R. Cass, seeks to address the links between international norm emergence and the translation of international norms into domestic political discourse and policy. Climate change provides opportunities to study domestic norm salience, as climate policy touches upon a number of important policy areas, including energy, transportation, commerce, taxation and foreign policy. This chapter presents an eight point scale to measure the domestic salience of an international norm. The scale is then utilized to measure the evolution of the international norm requiring domestic emission reduction commitments in the United States, Germany and the United Kingdom between 1985 and today.

Chapter 3, by Mary E. Pettenger, traces the formation of the Netherlands' climate change policy based on the Dutch definition and adoption of the norm of sustainable development. The Dutch case exemplifies the intricate dance between social values, such as economic growth and stewardship, alongside the material reality of threatening sea level rise. The norm of sustainable development, as conceptualized by the Dutch, embodies their unique understanding of climate change and environmental degradation.

Chapter 4, by Takashi Hattori, examines four Japanese domestic norms: economic growth, energy efficiency, international cooperation and environmental protection. These norms have framed and shaped Japan's climate change policy during three stages of Japanese policy development. The processes by which each norm has been defined, adopted and/or rejected demonstrate the struggle between power and knowledge as fundamental actors gained or lost power in Japan.

Finally, Part I concludes with Chapter 5, by Cathleen Fogel which begins to cross the divide between norms and discourses. This chapter examines the evolution of climate change norms and policies in the US in the early 2000s during the years of the G.W. Bush administration. Blocked from federal ratification of the Kyoto Protocol, domestic actors nevertheless transformed US climate change norms by pursuing strategies focusing on states, regions and cities. The norms of binding GHG reductions and increased mandatory use of renewable energy continued to spread throughout the US as religious communities, unions and corporations also expanded their climate discourses and commitments to justify climate protection action, including constructions emphasizing the economic benefits of climate policies, US energy independence, moral responsibilities and human health.

Part II illustrates the discourse analytical perspective. It begins with Chapter 6 by Karin Bäckstrand and Eva Lövbrand in which they present three discourses in the climate change arena: green governmentality, ecological modernization and civic environmentalism. This chapter serves as an important bridge between the norm-centered and discourse perspective in that it outlines the dominate discourses that shape the context from which international climate change norms are framed. At the same time, it provides an insightful discussion of how these three discourses have struggled over the last few years, perhaps leading to the emergence of the reform version of civic environmentalism as a new force in climate change negotiations and policy.

Matthew Paterson and Johannes Stripple challenge us, in Chapter 7, to be responsible in how we conceptualize agents and structures in climate change policy. They contend that the social construction of climate change has been produced principally through a discourse concerning territory and territorialization. While climate change is usually defined as a global issue to be dealt with collectively, Paterson and Stripple demonstrate how three discourses: threat/opportunity, emissions and equity, and carbon sinks, reframe climate politics in terms of reimposing territorial notions of political space. In addition, the chapter explores how these constructions at the same time are contradictory in a variety of ways.

Chapter 8 by Myanna Lahsen delves deeper into the construction of climate change knowledge and marginalized voices in the Southern state of Brazil. Her chapter focuses on knowledge creation, science and power, and cuts to the heart of potential successes and failures of climate change policy formation, adoption and implementation. Science, perceived as the search for objective knowledge, has been pushed and shoved by many who want to employ it in their quest for power. Distrust of this knowledge formation by Southern decision makers has led to suspicion of international climate policy and may impede future policy formation.

Heather A. Smith, in Chapter 9, draws from critical and feminist theory to explore how indigenous voices provide a counterhegemonic discourse to the "global" Western social construction of climate change. The chapter begins by identifying the mechanisms by which the global construction of climate change in the dominant discourse has been constructed. In the heart of the chapter, she provides a forum for declarations by indigenous peoples made at UNFCCC COP meetings from 1998 to 2004 which counter universalized claims of the global nature of climate change.

Finally, William D. Smith, in Chapter 10, offers us a cosmological discourse in his case study of a Totonac smallholder farming community in east central Mexico with a distinct land ethic, conception of community, and interpretation of climate issues. The chapter asserts that exploring understandings of climate change cross-culturally allows fresh thinking concerning relationships between humans and the rest of nature, about who plays key roles in climate change, and about the nature of responsibility. In particular, an analysis of the Totonac notion of "presence of mind" demonstrates an *intersubjective* approach to relationships between humans and various classes of non-humans, an alternative to the nature-society dualism that marks Western frameworks.

In the Conclusion, Loren R. Cass and Mary E. Pettenger return to the overall theme of power and knowledge to explore dominant and marginalized agents, structures and processes in the social construction of climate change, and to the themes raised in the Introduction. The reader is encouraged to think carefully about the interaction of power (who has power and for what purposes) and knowledge (what is knowledge, how is it defined and created) while reading the book. Additionally, attention should be paid as the book engages in a dialogue with constructivism, in its own recursive process, through the lens of climate change to bring greater understanding of the approach.

In conclusion, for over twenty years, individuals, international organizations and states have issued a clarion call to the world to confront the threat of anthropogenic climate change; however, little has been done to effectively address the issue. How has such an important issue not brought profound and substantially effective international action? Questioning the effectiveness and proper composition of international policies is a Sisyphean task for practitioners and academics. In essence, this topic embodies the core of the international politics perspective, seeking to uncover the hidden processes of politics by which people organize themselves to achieve what they desire, and how power, order and justice (or injustice) are revealed in these practices.

References

Adger, W.N., Paavola, J., Huq, S. and Mace, M.J. (2006), *Fairness in Adaptation to Climate Change*. Cambridge, MA: The MIT Press.

Adler, E. (2002), "Constructivism and International Relations" in W. Carlsnaes, T. Risse and B.A. Simmons, eds. *Handbook of International Relations*. London: Sage Publications, pp. 95–118.

________. (1997), Seizing the Middle Ground: Constructivism in World Politics. *European Journal of International Relations*, 3 (3), pp. 319–363.

Bäckstrand, K. (2004), "Civic Science for Sustainability: Reframing the Role of Scientific Experts, Policy-makers and Citizens in Environmental Governance" in F. Biermann, S. Campe and K. Jacob, eds. *Proceedings of the 2002 Berlin Conference on the Human Dimensions of Global Environmental Change "Knowledge for Sustainability Transitions. The Challenge for Social Science"*. Global Governance Project: Amsterdam, Berlin, Potsdam and Oldenburg, pp. 165–174.

Bauer, J. ed. (2006), *Forging Environmentalism: Justice, Livelihood, and Contested Environments*. New York: M.E. Sharpe.

Björkdahl, A. (2002), Norms in International Relations: Some Conceptual and Methodological Reflections. *Cambridge Review of International Affairs*, 15 (1), pp.9–23.

Blue Earth Alliance (2006), *World View of Global Warming: Glaciers, Glacial Warming, Receding Glaciers*. <http://www.worldviewofglobalwarming.org/pages/glaciers.html>.

Broadhead, L.A. (2002), *International Environmental Politics: the Limits of Green Diplomacy*. Lynne Rienner.

Burch, K. (2002), "Toward a Constructivist Comparative Politics" in D. Green, ed. *Constructivism and Comparative Politics*. New York: M.E. Sharpe, pp. 60–87.

Cass, L.R. (2006), *The Failures of American and European Climate Policy: International Norms, Domestic Politics, and Unachievable Commitments*. Albany, New York: State University of New York Press.

Checkel, J. (2006), Tracing Causal Mechanisms. *International Studies Review*, 8 (2), pp. 362–370.

________. (2005), International Institutions and Socialization in Europe: Introduction and Framework. *International Organization*, 59 (Fall), pp. 801–826.

________. (1997), International Norms and Domestic Politics: Bridging the Rationalist-Constructivist Divide. *European Journal of International Relations*, 3 (4), pp. 473–495.

Conca, K. and Dabelko, G.D. (2004), *Green Planet Blues: Environmental Politics from Stockholm to Johannesburg*, 3rd ed. Boulder, Colorado: Westview Press.

Cortell, A.P. and Davis Jr., J.W. (2000), Understanding the Domestic Impact of International Norms: A Research Agenda. *International Studies Review*, 2 (1), pp. 65–87.

Crichton, M. (2004), *State of Fear*. New York: Harper Collins.

Debrix, F., ed. (2003), *Language, Agency, and Politics in a Constructed World*. New York: M.E. Sharpe, Inc.

Demeritt, D. (2001), The Construction of Global Warming and the Politics of Science. *Annals of the Association of American Geographers*, 91 (2), pp. 307–337.

Dessler, A.E., and Parson, E.A. (2006), *The Science and Politics of Global Climate Change: A Guide to the Debate*. Cambridge, UK: Cambridge University Press.

Doty, R. (1993), Foreign Policy as Social Construction: A Post-Positivist Analysis of US Counterinsurgency Policy in the Philippines. *International Studies Quarterly*, 37 (3), pp. 297–320.

Dunn, K. (2006), Examining Historical Representations. *International Studies Review*, 8 (2), pp. 370–381.

Eder, K. (1996), *The Social Construction of Nature*. London: Sage Publications.

Emmerich, R. (2004), *The Day After Tomorrow*. Twentieth Century Fox.

Farrell, T. (2002), Constructivism Security Studies: Portrait of a Research Program. *International Studies Review*, 4 (1: Spring), pp. 49–72.

Faure, M., Gupta, J. and Nentjes, A. (2003), *Climate Change and the Kyoto Protocol*. Cheltenham, U.K.: Edward Elgar.

Fierke, K. (2001), "Critical Methodology and Constructivism" in K. Fierke and K. Jørgensen, eds. *Constructing International Relations: the next generation*. New York: M.E. Sharpe, pp. 115–135.

Fierke, K. and Jørgensen, K.E., eds. (2001), *Constructing International Relations: the next generation*. New York: M.E. Sharpe.

Finnemore, M. (1996), *National interests in international society*. Ithaca, NY: Cornell University Press.

Finnemore, M. and Sikkink, K. (2001), Taking Stock: The Constructivist Research Program in International Relations and Comparative Politics. *Annual Review of Political Science*, 4, pp. 391–416.

Fisher, D.R. (2004), *National Governance and the Global Climate Change Regime*. Latham, MD: Rowman & Littlefield.

Green, D., ed. (2002), "Constructivist Comparative Politics: Foundations and Framework" in *Constructivism and Comparative Politics*. New York: M.E. Sharpe, pp. 3–59.

Guggenheim, D. (2006), *An Inconvenient Truth*. Paramount Classics and Participant Productions.

Haas, P.M. (2004), When does power listen to truth? A constructivist approach to the policy process. *Journal of European Policy*, 11 (4: August), pp. 569–592.

Hajer, M.A. (1995),*The Politics of Environmental Discourse: Ecological Modernization and the Policy Process*. Oxford: Clarendon Press.

Hannigan, J. (1995), *Environmental Sociology*. New York: Routledge.

Hay, C. (2001), What Place for Ideas in the Structure-Agency Debate? Globalisation as a 'Process Without a Subject'. *First Press: Writing in the Critical Social Sciences*. <http://www.theglobalsite.ac.uk/>.

Hoffmann, M.J. (2005), *Ozone Depletion and Climate Change*. Albany, New York: State University of New York Press.

Hollis, M. (1994), *The Philosophy of Social Science: An Introduction*. Cambridge, UK: Cambridge University Press.

Hopf, T. (2002), *Social Construction of International Politics: Identities and Foreign Policies, Moscow, 1955 and 1999*. Ithaca, NY: Cornell University Press.

Hotz, R.L. (2006), Wildfire Increase Linked to Climate. *Los Angeles Times*, (July 7). reprinted by <http://www.stopglobalwarming.org/sgw_read.asp?id=544067102006> [accessed December 14].

Inayatullah, N. and Blaney, D. (2004), *International Relations and the Problem of Difference*. New York: Routledge.

Jasanoff, S. and Martello M.L. (2004), *Earthly Politics: Local and Global in Environmental Governance*, Cambridge, MA: The MIT Press.

Katzenstein, P., ed. (1996), *The culture of national security: norms and identity in world politics*. New York: Columbia University Press.

Klotz, A. and Lynch, C. (2006), Translating Terminologies. *International Studies Review*, 8 (2), pp. 356–362.

Kratochwil, F.V. (1989), *Rule, Norms and Decisions: On the conditions of practical and legal reasoning in international relations and domestic affairs*. Cambridge: Cambridge University Press.

Kubálková, V., N. Onuf and Kowert, P., eds. (1998), *International Relations in a Constructed World*. New York: M.E. Sharpe, Inc.

Lahsen, M. (2005), Technocracy, Democracy, and US Climate Politics: the Need for Demarcations. *Science, Technology, & Human Values*, 30 (1), pp. 137–169.

Litfin, K.T. ed. (1998) *The Greening of Sovereignty in World Politics*. Cambridge, MA: The MIT Press.

________. (1994), *Ozone Discourses: Science and Politics in Global Environmental Cooperation*. New York: Columbia University Press.

Livesey, S. (2002), Global Warming Wars: Rhetorical and Discourse Analytic Approaches to ExxonMobil's Corporate Public Discourse. *The Journal of Business Communication*, 39 (1: January), pp. 117–148.

Luterbacher, U. and Sprinz, D., eds. (2001), *International Relations and Global Climate Change*. Cambridge, MA: The MIT Press.

Miller, C.A. and Edwards, P.N. (2001), *Changing the Atmosphere: Expert Knowledge and Environmental Governance*. Cambridge, MA: The MIT Press.

Milliken, J. (2001), “Discourse Study: Bringing Rigor to Critical Theory” in D. Green, ed. *Constructivism and Comparative Politics*. New York: M.E. Sharpe.

________. (1999), The Study of Discourse in International Relations: A Critique of Research and Methods. *European Journal of International Relations*, p. 5.

NASA (2006), *2005 Warmest Year in Over a Century*. L. Jenner, ed. <http://www.nasa.gov/vision/earth/environment/2005_warmest.html>.

Newell, P. (2000), *Climate for Change: Non-state Actors and the Global Politics of the Greenhouse*. Cambridge: Cambridge University Press.

Oels, A. (2005), Rendering Climate Change Governable: From Biopower to Advanced Liberal Government? *Journal of Environmental Policy & Planning*, 7 (3: September), pp. 185–207.

Onuf, N.G. (2006), Private email correspondence with Mary Pettenger, November 27.

________. (1989), *World of our making: rules and rule in social theory and international relations*. Columbia: University of South Carolina Press.

Oreskes, N. (2004), Undeniable Global Warming. *Washington Post*, (December 26), p.B07. <http://www.washingtonpost.com/wp-dyn/articles/A26065-2004Dec25.html> [accessed December 14, 2006].

Paterson, M. (2001), Climate Policy as accumulation Strategy: the Failure of COP6 and Emerging Trends in Climate Politics. *Global Environmental Politics*, 1 (2: May), pp. 10–17.

________. (1996), *Global Warming and Global Politics*. London: Routledge.

Payne, R.A. (2001), Persuasion, Frames and Norm Construction. *European Journal of International Relations*, 7 (1), pp. 37–61.

Reus-Smith, C. (2005), “Constructivism” in *Theories of International Relations*, 3rd edition. New York: Palgrave Macmillan, pp. 188–212.

Rosa, E.A. and Dietz, T. (1998), Climate Change and Society: Speculation, Construction and Scientific Investigation. *International Sociology*, 13 (4: Dec), pp. 421–455.

Rowlands, I.H. (1995), *The Politics of Global Atmospheric Change*. Manchester: Manchester University Press.

Ruggie, J. (1998), *Constructing the World Polity: Essays on International Institutionalization*. New York: Routledge.

Schlesinger, J. 2005, “The Theology of Global Warming” *Wall Street Journal* August 8, A10.

Spaargaren, G., Mol, A.P.J. and Buttel, F.H. (2006), *Governing Environmental Flows: Global Challenges to Social Theory*. Cambridge, MA: The MIT Press.

Steinberg, P.E. (2001), *The Social Construction of the Ocean*. Cambridge: Cambridge University Press.

Toepfer, K. (2005), as quoted in *Kyoto Protocol About to Bite, UN Calls It First Step*. <http://www.planetark.com/dailynewsstory.cfm/newsid/29546/newsDate/16-Feb-2005/story.htm>.

Weldes, J. (1998), Bureaucratic Politics: A Critical Constructivist Assessment. *Mershon International Studies Review*, 42 (2), pp. 216–225.

Wendt, A. (1999), *Social Theory of International Politics*. Cambridge: Cambridge University Press.

________. (1992), Anarchy is what states make of it: the social construction of power politics. *International Organization*, 46, pp. 395–421.

Zehfuss, M. (2002), *Constructivism in International Relations: The Politics of Reality*. Cambridge: Cambridge University Press.

________. (2001), "Constructivisms in International Relations: Wendt, Onuf and Kratochwil" in K.M. Fierke and K.E. Jørgensen, eds. *Constructing International Relations: the Next Generation*. New York: M.E. Sharpe, pp. 54–75.

PART I
Norm-Centered Perspective

Chapter 2

Measuring the Domestic Salience of International Environmental Norms: Climate Change Norms in American, German, and British Climate Policy Debates

Loren R. Cass

This chapter focuses on the early framing of climate change as a political problem at the international level and the associated normative debates related to how states should respond. It then evaluates how this process was translated into the domestic political processes of the United States, Germany, and the United Kingdom. Early domestic framing of the problem had profound effects on the political debates in all three countries and played a major role in shaping the degree to which emergent international norms were translated into the domestic policy. Germany, and later the United Kingdom, accepted emission reduction commitments and pursued domestic greenhouse gas (GHG) emission reduction policies that in many cases could be difficult and costly to achieve. In contrast, the United States denied that climate change was sufficiently well understood to justify costly domestic policy changes. What explains the differences in national responses? Considerations of domestic economic competitiveness and resource transfers resulting from climate policy appear to have shaped the parameters within which the normative debates occurred; however, the logic of appropriateness and associated normative debates in each country placed significant limits on the pursuit of material objectives.

The literature on international norms (defined as collective expectations about the proper behavior for a given actor) has tended to focus on the role of persuasion and social learning among political leaders to explain norm emergence (Jepperson, Wendt, and Katzenstein 1996, 54). However, this chapter builds upon the work of Cortell and Davis (2000, 1996), as well as Klotz (1995) and Gurowitz (1999) in focusing on the relationship between domestic politics and international norms. Most environmental problems necessitate domestic legislative and regulatory reform to address the problem, which typically requires support beyond senior political leaders. Normative debates must penetrate more fully into the domestic political dialogue to influence national policy. This chapter seeks to evaluate the shifting normative framing of the problem at the international and domestic levels in the United States, Germany, and the United Kingdom to understand how normative debates and policy processes interact to alter state policy and behavior.

The Domestic Political Salience of International Norms

The focus of this study is the relationship between international norms and domestic policy rather than the process of international norm emergence. International norms develop concurrently with domestic and foreign policy formulation. This study seeks a deeper understanding of the interplay of international norms and domestic policy. Which norms will be incorporated into domestic policy and why? When is a norm likely to influence the formulation of domestic and foreign policy responses to climate change? These questions point to the problem of measuring domestic norm salience, which refers to the norm's level of influence in domestic political dialogue (Cortell and Davis 2000, 68). To what extent does an international norm constrain national behavior or create obligations for action? To what extent do political actors appeal to the norm to justify domestic policies or to block policy changes? In other words, how influential is the norm in shaping national political dialogue and behavior?

Domestic political rhetoric provides a starting point for evaluating norm salience (Cortell and Davis 2000). Rhetorical norm affirmation provides early evidence of the acknowledgement of an emergent international norm or support for a preferred alternative norm that may eventually be incorporated into domestic and foreign policy positions. However, political leaders may also cynically manipulate rhetorical norm affirmation to deflect political pressure and avoid concrete action. The combination of behavior and rhetoric provides greater evidence to evaluate the domestic salience of the emergent norm.

This chapter focuses on policies to address climate change in the fields of energy, transportation, commerce, taxation, and foreign policy. The politics of these policy areas can be highly contentious and are intimately linked to issues of economic competitiveness, economic growth, and domestic standards of living. These links provide opportunities to analyze the relationships among international norms, national rhetoric, international negotiating stances, and domestic climate policies.

Most constructivist scholars emphasize social diffusion and persuasive communication in altering the intersubjective understanding of the proper response to a given set of circumstances as the central mechanisms in norm emergence (Kowert and Legro 1996). Actors attempt to "frame" normative ideas in a way that resonates with existing norms and with the interests of the target audience. Frames are "specific metaphors, symbolic representations, and cognitive cues used to render or cast behavior and events in an evaluative mode and to suggest alternative modes of action" (Barnett 1999, 15). Frames are tools used to define a problem, to mobilize support for a particular response, and to persuade a target audience of the appropriateness of a proposed normative response. However, the emphasis on persuasion often obscures the important role played by material factors (Payne

2001; Checkel 1999; Finnemore and Sikkink 1998). Actors strategically use norms to pursue both ideational and material interests. At the same time, actors may create incentive structures to influence norm acceptance and compliance without redefining the target's preferences. Thus, norms may emerge that do not necessarily reflect the beliefs and preferences of most actors but rather reveal calculated norm compliance to achieve benefits and avoid costs.

In addition, coercion often plays a prominent role in norm affirmation, which suggests that rhetoric and even behavior may not reflect true motivations. Actors respond to shifts in incentive structures created by supporters and opponents of the nascent norm. If behavior is dictated by coercion, a state will be more likely to alter its behavior when the consequences of compliance or noncompliance change. On the other hand, if the actor is persuaded that a norm is appropriate, then it is more likely to be transcribed into the domestic political dialogue, and compliance is probable regardless of changes in incentive structures. This issue is especially important during the early phases of norm emergence. Once a norm becomes institutionalized domestically—even if it is the result of coercion—it is likely to continue to influence national behavior regardless of shifts in the underlying incentive structure (Finnemore and Sikkink 1998).

The focus on the effect of international norms on state behavior raises a larger question related to the target actors that must be persuaded or coerced to accept a norm. The constructivist literature has emphasized the role of two primary forces shaping elite support for an emerging international norm (Checkel 1999, 1997). First, norm entrepreneurs may mobilize international and domestic support for a norm to coerce political leaders to affirm the norm either by threatening political consequences or by "shaming" through concerted efforts to condemn national leaders for their failure to accept the norm (Risse 2000). A second mechanism involves a process of learning on the part of national leaders, who are persuaded of the appropriateness of the norm and accept the behavioral imperatives associated with the norm. In this case, the norm is likely to have maximum effect on the definition of interests and the design of specific policies. It is likely that both mechanisms will affect the process of norm affirmation. Some leaders may be persuaded of the appropriateness of the norm, and others may respond to coercion. Different combinations of persuasion and coercion will likely be at work in different countries. Domestic institutional variation may also affect the ability of norm entrepreneurs to influence national leaders (Checkel 1999).

Acceptance of an international norm among powerful decision makers is necessary, but not sufficient, for the norm to alter state behavior. Most policy areas involve multiple linked issues and much larger groups of participants. The domestic implementation of the policy requirements associated with a norm may require both the support of national leaders and at least the passive support or ambivalence of a large number of other actors. Climate change is a good example of this type of situation. The executive has significant leeway to conduct foreign policy and negotiate on behalf of the country; however, addressing climate change entails domestic policy changes that require broad political support to enact them.

For example, the norm requiring states to adopt a domestic GHG emission reduction strategy requires changes in energy, industrial, taxation, and/or commercial policy. The implementation of the behavioral imperatives contained in the emergent norm requires the persuasion or coercion of a broad set of domestic actors for the state to effectively implement the policies associated with the norm.

Domestic Salience Scale

The chapter now turns to describe an eight point scale included below to measure domestic norm salience in order to analyze the likelihood of an international norm altering national behavior. Domestic norm salience can be arrayed along a continuum from domestic irrelevance to a "taken for granted" status where the norm is embedded in domestic institutions and policies (Cortell and Davis 2000). The scale builds upon Cortell and Davis' (2000) three level scale of salience, which focuses on the domestic salience across "domestic actors." Significantly, however, the scale below differentiates between the salience of the norm for the domestic political leadership and broader public discourse. Political leaders may accept a nascent international norm and seek to incorporate it into domestic policy, or reject the norm. If it is rejected, it is possible for the norm to become embedded in domestic political discourse by resonating with important domestic actors, which may then pressure the leadership to affirm the norm and act on it. It is thus necessary to differentiate among the political leadership and other important domestic actors as well as between norm affirmation as the result of persuasion of the appropriateness of the norm and norm affirmation resulting from coercion. The scale below attempts to capture the effects of these variables on norm affirmation.

1. *Irrelevance:* National leaders do not acknowledge the emergent international norm in any way, and it is not a part of the foreign or domestic policy dialogue. National leaders do not even feel compelled to justify actions that contravene the proposed norm.
2. *Rejection:* National leaders acknowledge a proposed norm but reject it. The state will likely support an alternative norm and engage in debate with supporters of the less desirable alternative. The dialogue is conducted primarily on the international stage, and the normative debate has not entered mainstream domestic political dialogue.
3. *Domestic Relevance:* National leaders continue to reject the proposed international norm, but it has entered the domestic political dialogue. At this point the government faces pressure from both international and domestic actors to affirm the emergent norm.
4. *Rhetorical Affirmation:* National leaders affirm the international norm as a result of political pressure from within and/or internationally. The norm is now a part of the domestic and foreign policy dialogue, but it has not been translated into foreign or domestic policy changes.

5. *Foreign Policy Impact:* National leaders adjust the state's foreign policy to affirm the norm and may support its inclusion in international agreements. The change in position may be the result of persuasion of the appropriateness of the emergent norm or through domestic and/or international coercion. However, national leaders continue to reject changes in domestic policy to implement the norm's behavioral imperatives, or domestic actors continue to reject the norm and block domestic changes required by the norm.
6. *Domestic Policy Impact:* National leaders and other actors begin to justify changes in domestic policy on the basis of the international norm. At this point, the policy changes typically serve other purposes as well, but the norm provides additional justification for the changes. The norm is fully embedded in the domestic political dialogue, but the onus is still on the supporters of the norm to justify policy changes that may adversely affect domestic interest groups.
7. *Norm Prominence:* Domestic interest groups that wish to continue policies or pursue new initiatives that contradict the norm must now justify the violation of the norm. The burden of proof has shifted and the norm is becoming embedded in the domestic institutional structures and policies of the state.
8. *Taken for Granted:* The norm has become embedded in the domestic institutional structure of the state, and compliance with the norm is nearly automatic (Cass 2006, 9).

This scale will be employed to evaluate the process by which international norms achieve domestic political salience and impact domestic policy debates. Particular attention will be paid to the relative importance of ideational and material forces shaping the degree to which international norms achieve domestic political salience. The study turns now to describing one significant international norm that has influenced domestic climate policies in the three countries under study.

Measuring the Domestic Salience of the CO_2 Emission Reduction Commitment Norm

By the mid-1980s many scientists had become convinced that anthropogenic climate change was occurring and that a multilateral international response was necessary to address the threat. The scientists, operating through the World Meteorological Organization (WMO) and the United Nations Environment Program (UNEP), played a significant role in "framing" climate change for political debate at two levels. First, scientists were largely responsible for the initial framing of the problem for diplomats, but this framing then had to be translated into the domestic political context of each country.

As a consensus began to emerge that anthropogenic climate change was occurring, policy makers began to focus on the appropriate political response. Beginning in the mid-1980s, norm entrepreneurs attempted to build support for a norm requiring developed states to accept a CO_2 emission reduction commitment by arguing that developed states were historically responsible for the vast majority of CO_2 emissions and should bear initial responsibility for reducing global emissions. Calls for states to commit to quantitative emission reduction targets became a staple of environmental NGOs, who sought to pressure states to demonstrate their concern for addressing climate change. The effect of the international CO_2 emission reduction commitment norm led to significantly different responses in the three cases, and the domestic salience of the norm has varied over time. The case studies will explore the variance in domestic salience across countries and over time within each country.

The United States

By the mid-1980s, the American government had demonstrated an interest in the potential threat of climate change, but the issue had not yet been framed for political debate. Public awareness increased rapidly during the 1980s. A number of policy entrepreneurs seized upon climate change as a potentially useful issue—particularly in Congress. On-going negotiations regarding ozone depletion and its predicted effects on the health of the American people generated additional interest in the relationship between human activity and the atmosphere. In addition, a harsh drought in the upper Midwest and Canada, record high temperatures throughout the summer of 1988, and a dramatic fire in Yellowstone National Park combined to raise the public profile of climate change.

In response to the growing public interest, the Reagan administration advocated the continued study of climate change, but it consistently rejected calls for a policy response as premature. The administration's views on climate change were spelled out by Acting Assistant Secretary of State, Richard J. Smith, in testimony before a House Committee in 1988. He asserted that it would be "premature" to seek an international convention on climate change. "We believe it is time for governments to take a hard look at what we know and what we need to know about climate change and its potential for impacts in order to provide governments with a sound consensus of scientific evidence from which policy options can be developed" (US Congress 1988, 42). If scientific evidence eventually demonstrated that climate change was a significant danger to the United States, then the government would initiate appropriate responses. In addition, the administration asserted that emerging technologies would produce larger reductions in CO_2 emissions at a much lower cost in the future. It was therefore prudent to wait for a more complete scientific analysis of the threat and for more cost-effective technologies to address the problem. The Reagan administration framed climate change as a scientifically uncertain, long-term problem that was worthy of additional study. In addition, it portrayed policy initiatives to reduce GHG emissions as excessively costly and scientifically unjustified in the short term. Economic costs were an integral part of the climate debate in the United States from the beginning; whereas, most other states focused on the science of climate change, and cost issues emerged later in the debate. This framing of the problem was incompatible with a norm requiring an immediate commitment to reduce national CO_2 emissions. It would thus be necessary to alter the domestic framing of the problem before the norm could achieve significant salience.

The growing public interest in climate change inserted the issue into the 1988 presidential election. Republican candidate George Bush declared in a September 1988 speech in Boston that "those who think we're powerless to do anything about the greenhouse effect are forgetting about the White House effect. As President, I intend to do something about it" (Hecht and Tirpak 1995, 383). Candidate Bush's rhetoric gave renewed hope to those advocating stronger action, but the new Bush administration failed to pursue significant changes in American policy after taking office. A May 1989 dispute over the congressional testimony of NASA scientist James Hansen provided insight into the growing conflict within the Bush administration. Following Hansen's testimony before a Senate Committee, it was revealed that the Office of Management and Budget (OMB) had changed his testimony to make the prospects of climate change appear more uncertain. The revelation of OMB altering a scientist's testimony created a public outcry.

The Bush administration's position changed significantly following the dispute over Hansen's testimony. In the midst of the controversy, White House Chief of Staff John Sununu sent a telegram to the American delegation discussing climate issues in Geneva, instructing them to press for a global warming workshop to be held in Washington in the fall. The workshop was intended to provide a foundation for a "full international consensus on necessary steps to prepare for a formal treaty negotiating process" (Shabecoff May 1989, 1). The administration's shift in position followed intense bipartisan criticism from congress and environmental groups. The Bush administration had not suddenly become convinced of the merits of addressing climate change; rather, it faced a political incentive structure that demanded a shift in rhetoric. Privately, the White House acknowledged that the overwhelming criticism emanating from Congress and the public forced the administration to act (Shabecoff May 1989, 1). The norm requiring a CO_2 emission reduction commitment was circumventing the administration and directly entering the domestic political dialogue, which increased the pressure on the administration. The norm had achieved domestic relevance, level three on the salience index.

The November 1989 Noordwijk Conference on Global Climate Change provided a test of the new American openness to addressing climate change. The conference was structured as a stepping stone toward the negotiation of a climate convention. The United States faced significant pressure to accept the emerging norm requiring the freezing of CO_2 emissions by the year 2000. Only the United States, Japan, and the Soviet Union actively opposed the norm. The American delegation suggested that the United States might eventually agree to stabilization, but "[w]e would like to have a better understanding of the economic consequences" (Shabecoff July 1989, 9). The United States was one of the few countries raising concerns about the potential costs of climate policy. The American delegation rejected the proposed commitment, and the remaining states pledged themselves to a non-binding goal of stabilizing emissions by the year 2000, though the level at which they should be stabilized was left undefined. The norm requiring a domestic emission reduction commitment had achieved international prominence. However, the Bush administration continued to reject it, and there was insufficient domestic support to force the administration to alter its position.

The increasing international pressure for the US to fall in line and affirm the norm of freezing CO_2 emissions created a growing split within the administration. In February 1990, the IPCC was to meet in Washington, DC, and President Bush was scheduled to speak to the meeting. The speech split the Bush cabinet between those who argued for continued opposition to international action and those who argued that President Bush should use the opportunity of the IPCC meeting to seize a leadership role in climate policy. The president chose an intermediate course. He acknowledged that "human activities are changing the atmosphere in unexpected and unprecedented ways" (Weisskopf 1990, 1). This represented a shift in American rhetoric. Up to that point, the administration had consistently emphasized the uncertainty involved in climate science. He called for an "international bargain, a convergence between global environmental policy and global economic policy, where both perspectives benefit and neither is compromised .... Wherever possible, we believe that market mechanisms should be applied and that our policies must be consistent with economic growth and free-market principles in all countries" (Shabecoff Feb. 1990, 1). The President attempted to recast the normative debate. He acknowledged the international support for the domestic emission reduction commitment norm, but he rejected it and offered an alternative normative foundation for international climate policy. Emission reductions should be achieved globally utilizing the most cost-effective strategies. Rather than states assuming responsibility for reducing domestic emissions, he advocated economic efficiency on a global scale as the guiding principle for addressing climate change.

In April 1990 the White House sponsored the 'Conference on Science and Economic Research Related to Global Change' in an attempt to refocus the debate onto the uncertain costs of addressing climate change. However, the participants broadly condemned the conference as a stalling tactic. President Bush in his closing remarks to the conference responded to the storm of protest.

> We have never considered research a substitute for action .... to those who suggest that we're only trying to balance economic growth and environmental protection, I say they miss the point. We are calling for an entirely new way of thinking, to achieve both while compromising neither, by applying the power of the marketplace in the service of the environment (Shabecoff Apr. 1990, B4).

America's partners rejected the focus on economic costs and market mechanisms, and they continued to press for the United States to accept an emission reduction commitment.

There were opportunities in the early 1990s to rethink American climate policy. The Bush administration had launched a review of energy policy in 1989, which created an opening to reexamine the relationship between energy and climate policy. Congressional and presidential elections in the fall of 1992 provided an opportunity to debate American policy. However, the American economy weakened between 1990 and 1991. Job creation and economic growth dominated the 1992 elections. The American climate debate remained unsettled and embedded in conflicts over the science and economics of climate change. Beyond the energy strategy discussions, there was little substantive debate of domestic measures to reduce GHG emissions. Instead, the United States focused on achieving an international agreement that would be consistent with American preferences.

The Bush administration's negotiating position and domestic climate policy appeared to run counter to public opinion as the negotiations on a framework convention began in 1991. A poll conducted at the end of 1990 revealed that 69 percent of Americans believed that the US should join other countries in limiting emissions from fossil fuels. 73 percent of the respondents said they would be willing to spend more on fossil fuels to prevent serious effects from climate change (*ECO* 1991). The Bush policy also ran counter to the position of a small, but vocal, bipartisan group in congress that affirmed the emission reduction commitment. The Bush administration continued to reject the commitment norm, but the norm had entered the domestic political dialogue. The administration faced growing demands from domestic and international actors to adopt a CO_2 commitment. The United States, at least publicly, had become completely isolated. Every OECD country except the US and Turkey had agreed to a target and timetable for stabilizing or cutting emissions. It would be left to the new Clinton administration to re-evaluate the American position on the CO_2 emission reduction commitment.

The Clinton administration was much more sympathetic to international efforts to address climate change, but it faced strong domestic opposition to policies to reduce GHG emissions and a largely apathetic public. One of the new administration's first initiatives was an energy tax proposal, which it justified in part as a first step toward reducing American CO_2 emissions. The failure to pass the measure in a Democratic party controlled congress illustrates the lack of political salience that climate change had achieved in the United States. As a result, the administration largely accepted that policies to significantly reduce domestic emissions were politically impossible. The national CO_2 commitment norm remained relevant to the domestic climate debate, but domestic political forces were arrayed in a manner to block not only domestic initiatives to reduce emissions but even to block the affirmation of the norm by the government.

Echoing the Bush administration's efforts to reframe the international normative debate, the Clinton administration promoted "flexibility mechanisms" to achieve cost effective emission reductions on a global scale. The American flexibility proposals initially seized upon proposals to establish a program of Joint Implementation (JI), which would allow actors to invest in emission reduction projects in another country and receive credit for a portion of the reductions achieved. The JI proposal fit well with the American emphasis on market mechanisms to achieve emission reductions at the lowest possible cost. It also created the potential for the United States to avoid difficult domestic policies to reduce GHG emissions. Theoretically, JI should minimize the costs of reducing GHG emissions by equalizing the marginal costs of emission reductions across states.

The proposal met immediate resistance from NGOs who decried the attempt by the United States and other rich countries to purchase credits rather than reduce domestic emissions. Environmental NGOs sought to frame the JI debate as a question of governments taking legitimate domestic action to reduce GHG emissions versus crass attempts to purchase emission credits from abroad to avoid national responsibility.

In 1995 the Clinton administration finally bowed to international pressure and acknowledged that binding emission reduction commitments would be necessary. At the first meeting of the Conference of the Parties to the United Nations Framework Convention on Climate Change, the US officially accepted that a future protocol would contain an emission reduction commitment. However, it successfully began to alter the nature of that commitment. The commitment would cover multiple GHGs and not just CO_2. The US was able to achieve agreement that JI would be a part of the future protocol, and thus commitments could be met in part through "flexibility" provisions. The Clinton administration faced heavy domestic criticism for accepting the emission reduction commitment—particularly when the developing world would be exempt from the commitments. The administration had affirmed the norm that developed states should accept an emission reduction commitment, but the norm was not shared by a broad coalition within the American political system. The norm had achieved a "Foreign Policy Impact," level 5 on the salience scale. The constant international and domestic pressure created incentives for the administration to affirm the norm, but opposing domestic forces prevented policy changes to reduce American emissions.

The administration wanted to demonstrate international environmental leadership, and many individuals within the administration supported a commitment to reduce American GHG emissions. However, congress was hostile to international commitments that would alter American domestic policies. Industrial interests warned of the economic devastation that dramatic emission reductions would create. Finally, climate change did not resonate with the public. The administration stepped up its initiative to provide "flexibility" to efforts to meet national emission reduction commitments. A December 1996 position paper laid out the American proposals. "[T]argets could be set with budgets covering a multi-year period, for example, 2010–2020 ... . we should explore allowing Parties to borrow against their targets for the next period in order to emit more in a current period" (UNFCCC 1997, 37). Under the proposal, parties to the protocol would not have to reduce emissions for nearly twenty-five years, and even then, they could "borrow" emissions from a future period to avoid making reductions. The American paper also called for credits to be provided for joint implementation projects and for enhancing carbon sinks. European states and environmental NGOs broadly condemned the American position as undermining meaningful action to reduce GHG emissions. As "flexibility mechanisms" became more closely associated with American attempts to defer emission reductions, the norm requiring domestic commitments and policies to reduce emissions became further associated with legitimate action to address climate change.

The Clinton administration argued that the American experience with its sulfur trading program demonstrated the potential to achieve cost-effective emission reductions. The administration tried to frame trading as a legitimate and responsible approach to achieve emissions reductions. Secretary of Energy Federico Pena argued that emissions trading would "create incentives for countries with lower-cost emissions reductions to take these reductions and sell the 'extra' emissions allowances to those who face higher costs .... Global costs could be reduced by more than half .... As long as [emissions trading] is used to meet the world reduction goal, we see emissions trading as being entirely appropriate" (*International Environment Reporter* May 1997). The G-77, environmental NGOs, and the European Union (EU) framed the initiative as an American attempt to buy its way out of reducing domestic emissions. Dutch Environment Minister Margaretha De Boer, speaking as President of the European Environment Council, alluded to the difficulties that the United States faced in gaining acceptance for emissions trading. "[Emissions trading] might be fine for the US, but the cultures are different, and what will work for the US will not necessarily work for Europe" (*International Environment Reporter* June 1997). The EU argued consistently that national responsibility had to be the core principle rather than economic efficiency, as advocated by the US.

Up until the very end of the Kyoto Protocol negotiations, the United States rejected any commitment beyond a return of emissions to 1990 levels by the year 2012. The Clinton administration also attempted to introduce a number of loopholes and 'flexibility mechanisms' that would limit the actual emission reductions that the US would need to achieve. For the United States, the Kyoto negotiations were about designing an international regime to defer domestic action. Ultimately, the Clinton administration would accept a commitment to reduce US emissions by 7 percent below 1990 levels by the 2008–2012 commitment period. The administration had accepted the norm requiring a GHG emission reduction commitment in its foreign policy stance, but the norm had not achieved significant domestic policy salience. There was very little support in congress for policies designed to meet the commitment. In fact, the Senate made it clear that it would not ratify the Kyoto Protocol until at least the largest developing countries were forced to accept a GHG reduction commitment.

The victory of George W. Bush in the contested 2000 presidential elections produced substantial uncertainty both in American climate policy and in the international negotiations. The Bush campaign had called the Kyoto Protocol a flawed and unfair agreement, and it was unclear whether the new administration would be willing to continue the negotiations on implementing it. The international negotiations essentially stalled while the new administration contemplated its approach to climate policy. In March 2001, the administration declared that it considered Kyoto "dead." The administration's opposition to the protocol produced an avalanche of condemnation from environmental groups, Democrats, and foreign leaders. European leaders in particular were furious that the administration would so abruptly walk away from an agreement produced through years of arduous negotiations.

Despite the condemnations, the administration continued to reject the Kyoto Protocol and the norm requiring national emission reduction commitments. The administration asserted that it was willing to discuss international climate policy but only after emission reduction commitments were removed and an alternative normative framework could be established. The emission reduction commitment norm had failed to achieve significant domestic salience. It had slipped back to "Domestic Relevance," level three in the index of salience. It was part of the domestic political dialogue, but the administration and a large number of powerful interest groups continued to reject it. The American climate policy debate was never able to progress beyond the early framing of climate change as a scientifically uncertain, long-term threat that would be economically devastating to address. The result has been that the American debate continues to become bogged down in the science and economics of climate change, which has blocked the emergence of a foundation for policy action. Though national climate policy has remained stalled, there is increasing evidence of grassroots organizing and state and local government led climate initiatives that may provide the broad based political support to reframe the domestic climate debate, increase the domestic salience of international climate norms, and eventually force national political leaders to fulfill the policy obligations associated with the emergent norms (see Fogel's chapter in this volume and Rabe (2004) for a discussion of these ongoing changes).

Germany

Climate change was a particularly difficult issue for German political parties in the 1980s. The Social Democratic Party (SPD) had traditionally been viewed as more pro-environment than the Christian Democratic Union/Christian Social Union (CDU/CSU), but climate change split the party among anti-nuclear, environmental, and pro-coal factions. Addressing climate change fit well with the CDU's desire to promote nuclear energy, but it also had the potential to adversely affect the members of its business constituency. The emerging Green Party emphasized the elimination of nuclear power in Germany, which did not fit well with reducing CO_2 emissions if nuclear power was to be replaced by coal burning power plants. German climate policy was further complicated by the delegation of environmental responsibilities. The Federal Ministry for the Environment (BMU) was created in 1986, following the Chernobyl accident. Prior to its creation, environmental policy was split among various ministries. Climate policy fell under the purview of the Transport Ministry, which made an aggressive response to climate change unlikely.

As climate change emerged as an important international issue, parliament responded by creating the 'Inquiry Commission on Preventive Measures to Protect the Atmosphere' in December 1987 (Rest and Bleischwitz 1991). The formation of the inquiry commission offers an interesting comparison with the initial framing of the climate debate in the United States. The inquiry commission was composed of an equal number of scientists/policy experts and politicians. It was intended to produce an unbiased review of the relevant scientific evidence and, ideally, to provide non-partisan policy advice to parliament. The commission reviewed the existing scientific evidence, decided that there was credible evidence that climate change was occurring, and advocated preventative action to address a potentially serious threat. The Commission publicized its first report, "Protecting the Earth's Atmosphere; An International Challenge" in October 1988. The Commission called for a 30 percent CO_2 emission reduction target by 2005 (using a 1987 base year). It also argued that Germany could reduce its CO_2 emissions by 35 to 45 percent with existing technologies (Rest and Bleischwitz 1991, 161). The inquiry commission framed climate change as a serious threat that demanded both an international and a domestic response. Importantly, it was also a problem that could be addressed at minimal economic cost by using existing technologies. This framing of the problem marginalized opponents of CO_2 reductions and created the foundation for the government to pursue policies to reduce emissions.

Within this context, the Kohl government argued consistently that reductions in GHG emissions would be necessary to mitigate the adverse effects of climate change. In late 1989, Germany pushed at the Noordwijk climate conference for an agreement to freeze CO_2 emissions at 1989 levels by the year 2000. The United States, the USSR, and Japan opposed the German proposal, which forced Germany and its European partners to accept a weakened statement noting that "in the view of many industrialized nations ... stabilization of (CO_2) emissions should be achieved as a first step at the latest by the year 2000" (Abramson 1989). Germany emerged as one of the foremost advocates of establishing CO_2 emission reduction commitments for all developed states. The CO_2 commitment fit well with pre-existing German norms related to the role of government in taking a precautionary approach to environmental policy (Boehmer-Christiansen and Skea 1991; von Moltke 1991). The norm also did not appear to adversely affect important domestic actors. The norm thus rapidly achieved a "Foreign Policy Impact" (level five on the salience index) though it did not have an immediate impact on domestic policy.

By 1990, the Kohl government began to raise the prospect of introducing specific domestic policies to reduce GHG emissions. The government considered a number of proposals, including a carbon tax and energy efficiency regulations. Germany's ability to cut its CO_2 emissions would be given a substantial boost with reunification. The inefficient East German power industry and heavy industrial plants presented a wealth of opportunities to inexpensively reduce CO_2 emissions. The ease with which Germany could apparently cut its CO_2 emissions minimized the opposition from industrial interests and emboldened Germany's foreign policy position. Tough international targets would require little domestic policy reform, but they could force other states to undertake potentially expensive policy changes. German support for a CO_2 emission reduction commitment was thus consistent with the material interests of major economic actors.

The German government's activity on climate change also included an emphasis on research and development of alternative energies. In February 1990, the cabinet released its "Third Energy Research Program." The program focused largely on renewable energy sources and energy efficiency research. Nuclear energy made up the bulk of the renewables research, but wind and solar energy also received increased funding. By this point the CO_2 commitment norm had achieved a domestic policy impact (level six), and it had become much more integrated into the German political agenda than in either the British or American cases. The inquiry commission's conclusions that climate change was a serious threat and that domestic policy could sharply reduce GHG emissions at little additional cost marginalized arguments that the costs of GHG reductions were prohibitive. There was also no significant debate surrounding the basic science of climate change.

In April 1990 Environment Minister Klaus Toepfer announced that the government was formulating a plan to reduce Germany's CO_2 emissions by at least 25 percent by the year 2005. He asserted that "ecological necessities" must take precedence over long term economic development. Toepfer's willingness to concede that CO_2 reductions could be expensive and that they were still necessary sharply contrasted with the American position that only "no-regrets" policies were justified given the existing state of scientific knowledge (*Week in Germany* 1990). The momentum toward domestic policy actions to reduce CO_2 emissions created a split within the Kohl cabinet. The Federal Association of German Industry (BDI) had become concerned that the government was moving rashly toward costly policy measures that would damage the competitive position of German industry. The BDI in particular challenged the imposition of a climate tax as well as new regulations on industry. It argued that a cooperative approach involving voluntary efforts to reduce CO_2 emissions in energy production would be more efficient. Such concerns resonated within the cabinet. It was easy to pursue aggressive international commitments when there would be limited domestic economic effects, but the proposals for new policy measures increased the debate on the nature of the German response. The internal debates presented a test of the domestic salience of the CO_2 commitment norm. Despite the misgivings of industry, the cabinet announced the creation of an inter-ministerial working group to formulate proposals to meet the 25 percent target.

In the fall of 1990, the German government adopted a national CO_2 reduction program containing sixty measures to reduce CO_2 emissions. The measures included promoting combined heat and power generation, stronger energy efficiency standards, improved coal combustion technology, renewable energy development, a carbon tax, and increased vehicle efficiency. Toepfer announced that Germany had established itself as the leader in climate policy. "I may point out that, with this decision, we have clearly supported the leading position in the climate discussion worldwide" (Kabel 1990). The program was ambitious, but the measures were largely commitments to future action, and there was almost no discussion of reducing the country's reliance on coal for energy, which would have offered the greatest opportunities for reducing CO_2 emissions. Regardless, the CO_2 commitment had become firmly entrenched in the German policy process and was having a domestic policy impact (level six) as the government utilized its CO_2 commitment to justify significant changes in domestic policy.

The two inter-ministerial reports, released in November 1990 and December 1991, cataloged long lists of policies directed at reducing carbon emissions. Of these proposals, only one, a requirement for electricity companies to purchase power from renewable energy sources, had been fully adopted by the end of 1992. The Kohl government either delayed or rejected proposals to encourage combined heat/power cogeneration plants, to increase building insulation standards, to establish minimum heating efficiency standards, and to introduce a carbon and/or energy tax. The failure to enact these policies produced sharp criticism from the SPD, the Greens, and even Chancellor Kohl's own CDU/CSU party members. The government asserted that its new inter-ministerial report, due by the end of 1993, would contain sufficient measures to reduce all GHG emissions, including NOx, CFCs, VOCs, and CO_2 by 50 percent by 2005. The government's ambitious CO_2 reduction target was beginning to soften as the government diluted the focus on CO_2 and included additional GHGs in the calculations. It had also begun to incorporate the East German reductions into the national commitment despite earlier claims that the emission reductions would be achieved in the West.

Between the signing of the UNFCCC in Rio and the first Conference of the Parties (COP1) in 1995, climate policy intersected with related domestic policy debates in four primary areas: energy efficiency, the carbon tax, nuclear power, and transportation policy. In each case, the government declined to impose new regulations or taxes and instead pursued voluntary agreements with the affected actors. The voluntary approach fit well with Germany's corporatist system, and it was possible that the strategy could provide the necessary emission reductions at a lower cost than government regulation. Climate change remained a much more prominent issue in German politics than it was in the United States or the United Kingdom. The Green Party had been extremely active on climate issues, and the CDU had adopted climate change as one of its core 'green' issues in the most recent elections. The Kohl government continued to utilize climate change as a justification for expanding nuclear energy, but public, SPD, and Green Party opposition constrained the ability of the government to promote nuclear power. The government exploited climate change to justify a carbon/energy tax and additional energy efficiency measures, but in both cases, it ultimately rejected the measures on competitiveness grounds. It also refused to address emissions from transportation. Even though the emission reduction commitment remained popular in Germany, the norm was not sufficient to overcome the opposition of important economic actors. While the norm continued to have a domestic policy impact, it had not reached the seventh level of norm salience. Proponents of emission reductions continued to be forced to justify why emission reduction should take priority over economic interests. Material interests placed constraints on the policy effects of the norm.

Germany was one of the few countries to meet the UNFCCC 2000-stabilization target, and it would achieve short-term emission reductions greater than any of its industrial partners. However, the German reductions were primarily derived from the restructuring of the former East German economy. Emissions in the East were already showing signs of increasing. Germany had an interest in assuring that over the long-term it would be able to meet future GHG reduction goals without significant domestic costs. Thus, as the Kyoto meeting approached, Germany began to support American calls for including 'flexibility mechanisms', such as joint implementation, in a protocol to the FCCC. This represented a shift away from the norm requiring emission reductions to be achieved exclusively domestically. The Kohl government maintained a strong rhetorical position on climate change, and it continued to support a meaningful international agreement to reduce GHG emissions. However, it largely deferred costly domestic actions to reduce emissions and placed its faith in voluntary accords and emission reductions in the East to meet its commitments. The Kohl government claimed that Germany had already initiated more than 100 measures to address GHG emissions (Federal Republic of Germany 1994). The vast majority of these measures would have only minor effects on emissions, and it was becoming increasingly clear that they would be insufficient to meet Germany's domestic target.

The Kohl government was in a difficult political position. The economy was stagnating and unemployment levels remained unacceptably high. The government faced growing opposition to its 7.5 percent 'solidarity tax' that it had introduced to underwrite the costs of reunification. The budget deficit was increasing, which required spending cuts and increased tax revenues. The CDU/CSU position on environmental policy reflected growing public apprehension about the economy and decreasing concern for environmental affairs. The Kohl government altered the terms of the German environmental debate by raising issues of balancing economic costs and environmental benefits. The government commissioned a study in 1995 to examine the costs of meeting the 25 percent emissions reduction target. The study by two German research institutes concluded that meeting the interim target of cutting CO_2 emissions by 17 percent by 2000 would cost 120,000 jobs as a result of energy intensive industries scaling back their production and moving jobs overseas. In addition, 150,000 jobs would be sacrificed to meet the 25 percent target by 2005. According to the report, German industry would need to spend an additional 489.1 billion dollars to achieve the 25 percent reduction target by 2005 (*International Environment Reporter* 1996). German industries decried the absurdity of imposing the heavy costs of climate policy on them while foreign competitors faced no similar costs. German environmental organizations and opposition parties condemned the government's attempts to frame environmental policy as a choice between jobs and environmental quality, but economic interests were clearly constraining Germany's ability to reduce its GHG emissions.

The difficult domestic situation led the Kohl government to retreat from implementing an aggressive domestic GHG reduction policy. It maintained its commitment to cut emissions by 25 percent by 2005, but it refused to take significant additional measures to achieve the target. Germany continued to play a leadership role internationally, and the Kohl government attempted to force other states to undertake major new commitments to cut CO_2 emissions even as it scaled back its domestic initiatives. The government also began to pave the way for meeting its future commitments by affirming economic efficiency as a guiding principle of climate policy. The Kohl government became a supporter of JI and emissions trading—though it argued that these measures had to be supplementary to domestic emission reductions.

The SPD/Green victory in September 1998 produced a tumultuous period of reform in Germany that had the potential to significantly affect climate policy. Proposals to phase out nuclear power and to impose 'ecological taxes' dominated the early months of the new coalition government. In addition, the new government sought to devise mechanisms to promote renewable energy and combined heat and power generation (CHP). The SPD/Greens victory gave the ruling coalition control over both the Bundestag and the Bundesrat, and thus created a strong foundation for pursuing its agenda. The SPD/Green government faced a number of difficult tradeoffs related to climate policy. The most important issue for the Greens was the phasing out of nuclear energy. While the SPD supported an end to nuclear power, it also recognized that an immediate shutdown of nuclear plants would be politically costly as well as devastating to the country's CO_2 emission reduction commitments. Talks between the nuclear power industry and the government were extremely contentious and strained relations between the SPD and Greens. The proposed imposition of 'eco-taxes' also created dilemmas for the government. The creation of a carbon tax would provide the greatest incentives for reducing CO_2 emissions. However, such a tax would be ruinous to the SPD's traditional supporters in the coal industry. An energy tax would provide some incentives for energy efficiency but, to have a significant effect on CO_2 emissions; it would have to be relatively large, which would harm the competitive position of German industry. These were some of the dilemmas facing the Schroeder government as it began to chart its climate policy.

The government admitted that CO_2 emissions had only fallen by 15 percent by 2000. It would need to undertake significant additional measures if it was to achieve a 25 percent reduction by 2005. In April 2000, the Environment Ministry leaked a draft climate strategy to the press, which contained nearly three dozen measures designed to improve energy efficiency, lower transportation emissions, reduce industrial and energy transformation emissions, and produce the 100 million ton reduction in CO_2 emissions necessary to meet Germany's domestic commitment. The CDU and CSU blasted the plan. They argued that even if the government implemented all of the proposed measures, it would still fail to fulfill the national CO_2 commitment. Most of the reductions would be offset by increased emissions as nuclear power plants were closed and replaced by fossil fuels. Even within the Schroeder government, the proposal resulted in significant conflict.

The German public had become increasingly engaged in the climate policy debate. In a June 2000 public opinion poll, 98 percent of respondents perceived climate change to be an important problem. In addition, 58 percent said the government needed to do more to address the problem (Bundesministerium fur Umwelt 2000, 37). After extensive internal debate, the cabinet approved the climate strategy. The plan proposed a set of measures to cut German CO_2 emissions by 50 to 70 million tons by 2005. This was less than the 100 million tons contained in the earlier draft and reflected a reduction in the number of measures and estimates of achievable emission reductions, though it would theoretically provide the emission reductions necessary to meet the 25 percent target. The strategy also included commitments from 19 German industrial groups to increase their GHG emission reductions. In 1996 these groups had committed themselves to reducing their emissions by 20 percent by 2005. By 1999 they had already achieved aggregate CO_2 emission reductions of 23 percent. The groups committed themselves to cut their CO_2 emissions by 28 percent by 2005 and all GHG emissions by 35 percent by 2012 (*International Environment Reporter* 2000). The cabinet gave final approval to the strategy in October 2000.

On the international stage, Germany maintained its role as the foremost advocate of measures to address climate change. The Schroeder government continued to press for harmonized energy taxes at the EU level and maintained its pressure on the Umbrella Group to close perceived "loopholes" in the Kyoto agreement. The appointments of the Green Party's Jürgen Trittin, an uncompromising proponent of environmental protection, as Environment Minister and Joschka Fischer as Foreign Secretary enhanced the government's environmental image and led to a very aggressive approach to the international negotiations. The SPD/Green government faced a number of difficult decisions requiring tradeoffs among competing national interests. How fast and at what cost could the government accelerate national GHG emission reductions? What were the minimum requirements for an acceptable international agreement? These issues produced difficult debates both within the coalition government and between the coalition and the opposition. These debates reflected the prominent position of the German emission reduction commitment, but they also illustrated the economic constraints that shaped how those reductions would be achieved. The domestic emission reduction commitment has not achieved a "Taken for Granted" status, but it has clearly achieved "Norm Prominence" (level seven). The commitment and domestic policies to reduce emissions are broadly seen as appropriate and necessary despite their economic costs.

The United Kingdom

As climate change emerged as a significant political issue in the mid-1980s, the British response was heavily influenced by its recent experience with international environmental policy. The UK had been pilloried over its role in the acid rain problems of Europe. The Thatcher government's emphasis on deregulation and privatization had pushed environmental policy toward the bottom of the policy agenda. Environmental organizations castigated Britain as the "dirty man of Europe," and public pressure was building for the government to more aggressively address environmental issues. Prime Minister Thatcher shifted the Conservative Party stance on environmental issues in the fall of 1988 and adopted climate change as its highest priority environmental issue. The UK became a strong advocate for international negotiations on climate change, but it aggressively sought to control their evolution.

In April 1989, the British Foreign Office began actively promoting a climate convention as the most appropriate response to climate change. In May, the British permanent representative to the UN, Sir Crispin Tickell, called for the negotiation of a framework convention on climate change laying out principles for "good climatic behavior" (*International Environment Reporter* June 1989, 281). He argued that the convention should be completed as soon as possible with protocols to be negotiated later "as scientific evidence requires and permits" (ibid.). The British proposal put the UK at the forefront of international climate policy. The government supported a multilateral response; however, it was unwilling to take specific domestic actions, and the government had not yet acknowledged the CO_2 reduction commitment norm.

By the November 1989 Noordwijk ministerial conference on climate change, the international norm requiring states to adopt domestic CO_2 emission reduction commitments was achieving international prominence. The draft declaration called for signatory countries to commit themselves to the stabilization of CO_2 emissions by the year 2000 and for research into the possibility of reducing emissions by 20 percent by the year 2005. The UK was reluctant to go along with specific targets because it was wary of the potential costs to its economy. However, it also did not wish to be perceived as undermining the international negotiations. The government attempted to defer debates on emission reduction commitments. Junior Environment Minister, David Trippier, noted that any specific agreement would be "premature". "All countries should commit themselves to the work of the IPCC and not allow that work to be diverted at this stage by arbitrary and unsubstantiated targets. In any case, we have no basis for setting a target of 20 percent reductions when so many other figures have been cited elsewhere" (Neale and Wastell 1989, 4). Trippier instead called for all nations to support a framework convention with specific goals to be negotiated after the IPCC provided a scientific basis for action.

British negotiators acknowledged the CO_2 commitment norm, but they tried to alter the norm to fit British interests. Under a British compromise proposal, the signatories would commit to stabilize CO_2 emissions by the year 2000 with the levels of stabilization left undefined. However, even this was unacceptable to the United States, the USSR and Japan, and the final statement did not include the stabilization target. Labor leaders lambasted the government for undermining the possibility of a meaningful international agreement. "While other European countries are prepared to accept targets for reductions, our government and the USA. are refusing to join them. Doesn't this show the government's claim to be at the forefront in tackling global warming to be a bogus one" (*International Environment Reporter* Dec. 1989). The CO_2 commitment norm had rapidly achieved domestic relevance (level three). The government acknowledged the appropriateness of the norm, but it rejected the requirements entailed in the norm. However, it faced rising domestic and EU pressure to affirm the nascent norm.

In the midst of the Noordwijk conference, Prime Minister Thatcher addressed the United Nations General Assembly. She was put in the uncomfortable position of attempting to stress the importance of international environmental policy while at the same time justifying the UK's refusal to accept firm commitments to reduce CO_2 emissions. She stressed the active role that Britain had played in climate policy and promised to continue that role. She also argued that solutions to climate change had to involve "sound scientific analysis" before emission reduction targets could be negotiated (*International Environment Reporter* Dec. 1989). This fit well with British domestic policy norms that emphasized the importance of scientific evidence before pursuing significant policy changes. The influence of the precautionary principle was weaker in British environmental policy (Boehmer-Christiansen and Skea 1991). The focus on uncertain scientific evidence and concerns over economic consequences of emission reduction commitments created obstacles to acceptance of the emission reduction norm.

The Thatcher government sought to defer the targets discussion until after the completion of the IPCC report. Agreements to reduce CO_2 emissions could impair the planned privatization of the state-owned electricity industry. The electricity generation industry was one of the primary CO_2 emitters. The British government had already experienced the heavy costs to the industry in addressing sulfur emissions. The need to cut CO_2 emissions would likely create uncertainty surrounding the costs to the industry and undermine the proceeds from privatization. However, acceptance of an emission reduction commitment had emerged as the primary indicator of a country's support for addressing climate change. The government faced a dilemma. While it had accepted that climate change was an important issue, it was unwilling to accept the CO_2 commitment norm and internalize the domestic policy implications of its international commitment to addressing climate change, which weakened its claim to leadership on climate policy.

Within the EU, the UK faced increasing pressure to accept the CO_2 commitment. At a March 1990 EU Environment Council meeting, the Commission proposed that member states agree to an emissions freeze by 2000 and significant cuts by 2010. British Environment Minister Patten rejected emissions targets and called for stricter emissions limits on trucks and de-regulation of the transportation and energy sectors. He argued that action on these issues would reduce inefficiencies in the EU marketplace, could reduce CO_2 emissions by 100 million tons a year, and would be more effective than "simplistic political formulas" (Hunt 1990). The government was attempting to recast the normative debate. Rather than focusing on quantitative targets, governments should focus on specific policies to reduce GHG emissions. The British government was engaged in a debate over how the response to climate change should be framed. Would symbolic commitments to cut emissions be the benchmark for acceptance as a supporter of climate policy, or would specific policy initiatives be the focus? The government was loathe to accept quantitative emission reduction commitments when it was not clear how such targets could be achieved or how much it would cost to meet the commitments.

The impending release of the IPCC report in 1990 created pressure to accept a commitment to reduce GHG emissions. Britain had consistently argued that it would act when the IPCC had finished its work. Rather than attempt to undermine the IPCC's findings, Prime Minister Thatcher announced that the UK would agree to stabilize its CO_2 emissions at 1990 levels by the year 2005, but she qualified the commitment. "Provided others are ready to take their full share, Britain is prepared to set itself the demanding target of a reduction of up to 30 percent in presently projected levels of carbon dioxide emissions by the year 2005. This would mean returning emissions to their 1990 levels by that date" (Brown and Williams 1990). This was five years later than the other European countries had agreed and, according to the Prime Minister, would only occur if other countries (presumably the United States) acted as well. The UK affirmed the international norm requiring a quantitative emission reduction commitment, though it rejected the specific target date. The norm had become integrated into the UK's foreign policy (level five), but the government continued to balk at specific domestic measures to achieve it.

The advantage of 2005 for the British government was that most of the actions necessary to bring the UK in line with the EU's Large Combustion Plant Directive (LCPD) would take place between 1990 and 2003 with the majority occurring after 1998. Though directed at SO_2 and NO_X emissions, the LCPD created incentives for British electricity generators to switch from coal to natural gas for power generation to avoid expensive retrofitting of existing plants with flue gas desulphurization equipment. Natural gas plants were also cheaper to build and run. As older coal burning plants closed and newer, cleaner natural gas plants came on line, British emissions of CO_2 would naturally fall.

Following Margaret Thatcher's 1990 resignation, the new Major government faced rising domestic and EU pressure to bring its commitment into line with the international norm. The government attempted to capitalize on climate change to further its policy agenda. It linked the promotion of nuclear energy to its GHG reduction commitments. The government also desperately needed to increase government revenue to reduce the growing budget deficit. Climate change provided a justification for substantially increasing gasoline taxes. The CO_2 norm had achieved a domestic policy impact (level six). However, there were also significant limits to the policies that the government was willing to implement. The Major government sought to slow the conversion from coal to natural gas for power generation to protect the remnants of British Coal for privatization. It also continued to object to the proposed European carbon tax. The government faced a growing tension between its stated desire to fulfill its CO_2 emission reduction commitment and the potential adverse effects of reducing CO_2 emissions on domestic economic interests. It thus found it politically expedient to use the CO_2 commitment to justify the painful reforms that were necessary to achieve other objectives, but the government refused to pursue policies to reduce CO_2 emissions that would adversely affect important domestic constituencies such as the energy industry and industrial producers.

Despite the economic concerns, the new Major government adopted a novel strategy to increase public and business support for climate policy. In May 1991,

Environment Secretary Heseltine and Industry Secretary Peter Lilley set up a twenty-four member 'Advisory Committee on Business and the Environment' (ACBE), which was composed of the chief executives and directors of some of Britain's largest companies and chaired by Shell Oil chief executive John Collins. The committee was created to provide a framework for dialogue between industry and the government on environmental policy. The committee's first task was to recommend a set of policies to reduce British GHG emissions. The committee proposed a number of significant measures in its first report in September of 1991. It called on car companies to voluntarily improve gas mileage. It recommended higher gasoline taxes and taxes on less efficient automobiles. It also promoted more energy efficient consumer products through improved product standards and tax incentives. The group did not recommend any type of carbon or energy taxes; nor did it promote energy efficiency programs for industry. The report provided a foundation for the government's push for higher gas taxes. It also helped to build business support for additional measures to reduce GHG emissions.

In April, Prime Minister Major announced that the UK would move its target date for stabilization of carbon emissions from 2005 to 2000, in line with the rest of the European Union. He reiterated that the target remained contingent upon other countries (i.e. the United States) accepting a similar target. New emissions data suggested that the UK would likely meet the 2000 target without significant additional policies. The rapid shift from coal to natural gas was having a profound effect on British emissions. It was possible that under the most optimistic assumptions, the government would not have to undertake any new domestic policies to meet its objectives. Despite this, the Major government adopted a relatively proactive stance and assumed it would need to cut its emissions by an additional ten million tons of CO_2 by 2000 to return emissions to 1990 levels. It then prepared a program to achieve the emission reductions, which included an increase in the value-added taxes on home heating fuel, electricity, and transportation fuels. It also included energy efficiency investments, new energy efficiency building regulations, and support for renewable energies.

The government successfully pursued an increase in the gasoline tax and half of its proposed VAT on household energy use, but Tory backbenchers blocked the larger VAT increases. On the other hand, the government was unable to overcome obstacles to effectively implement its Energy Savings Trust program to promote energy efficiency and was unwilling to address industrial energy use through taxation due to competitiveness concerns. The government refused to tackle the transportation sector beyond raising the gasoline tax and declined to impose significant new regulations to improve energy efficiency. While the government proved to be much more aggressive in pursuing GHG emission reductions than the US, it failed to implement most of the proposals that would have the greatest effect on emissions. The Major government was willing to increase taxes on individuals but not on industry due to competitiveness concerns. Rather than pursue additional regulatory or taxation measures, the Major government emphasized voluntary energy conservation measures and publicity campaigns. It hoped that the broader structural changes occurring in the British economy would be sufficient to reduce emissions to 1990 levels by 2000. This approach heralded future problems for British climate policies. Fuel switching was a limited solution. To meet future reduction targets, the government would have to address other emission sources or accept international flexibility mechanisms such as Joint Implementation and emissions trading.

The Labour party made climate change a central component of its 1997 election campaign. In July, Labour declared that, if elected, it would reduce British CO_2 emissions by 20 percent by 2010. The pledge was much more ambitious than the Conservative's goal and reflected broad public support for addressing climate change. Following the election, the new Labour government faced a daunting task in its attempt to meet its impressive target. Prior to 1997, the switch from coal to natural gas for electricity production had produced the vast majority of British CO_2 emission reductions. There were limits to further fuel switching, and rapidly increasing transportation emissions threatened to undermine the progress that had been made. Achieving the 20 percent reduction target would require significant additional policy changes. Initially, the Blair government wavered in its support for the 20 percent target. It refused to make the target a legally binding part of the EU burden sharing agreement. An October 1998 draft climate change consultation paper asserted that the government's aim was to "start a national debate on how we can move beyond our legally binding target [the EU burden sharing target] towards a 20% reduction" (*ENDS Report* 1998). The paper also emphasized that while compliance with the Kyoto Protocol "cannot be jeopardized, we will not introduce measures that would damage competitiveness, nor will we take any action that would bring unacceptable social costs" (ibid.). The rhetoric fell considerably short of accepting a binding 20 percent target.

The government predicted that emissions would fall by approximately 8 percent by 2010. The government thus needed to achieve an additional 12 percent cut in emissions. The Labour government did not produce its strategy to achieve these reductions until November of 2000. The strategy emerged following a long and difficult consultation phase. The Labour government's climate strategy focused on four policy initiatives: a new climate levy on industrial energy use and associated voluntary agreements with industry to reduce GHG emissions, a transportation fuel duty escalator, renewable energy promotion, and transportation policy (Department of Environment 2000). All of these policy proposals proved to be politically difficult to enact and required many compromises. However, it is clear that the emission reduction commitment norm had become well established under the Labour government, and it would likely be difficult to reverse many of the associated policies in the future. The emission reduction commitment norm had achieved at least level six ("Domestic Policy Impact") and is on its way toward norm prominence.

Conclusion

Clearly, the domestic political salience of the domestic emission reduction commitment norm varied widely across and within the cases over time. In all three states, the political leadership eventually affirmed the international norm that states should accept a GHG emission reduction commitment, though the second Bush administration would ultimately reject it. The first conclusion that flows from the case studies is that the public perception of a norm is critical to its salience. The perception of the emission reduction commitment norm as being both effective and fair made it difficult to argue against its legitimacy. The public accepted the position of norm entrepreneurs presented in the introduction: industrial states were historically responsible for the vast majority of GHG emissions and should bear the primary responsibility for reducing global emissions. In addition, an emission reduction commitment became the symbol of a country's support for, or opposition to, addressing climate change (for a further discussion of framing and perception in the social construction of climate change, see Paterson and Stripple in this volume).

However, in each country, the speed with which the norm achieved domestic salience was significantly affected by pre-existing political norms and the initial framing of the problem. Germany's environmental policy emphasis on the precautionary principle and the framing of the problem by the inquiry commission created an environment conducive for the norm to achieve rapid political salience. Not only did the government promptly affirm the norm internationally, but it also quickly began to justify domestic policy changes on the grounds that they were necessary to achieve emission reductions to fulfill Germany's international obligations. However, even in this case, it is clear that consideration of German competitiveness and the costs of emission reductions for energy companies placed limitations on the pursuit of emission reductions.

In the United Kingdom, the emission reduction commitment did not initially resonate with either the political leadership or with the public. The government accepted that climate change was a potential threat, and it sought to address it. However, it was reluctant to take on specific policy commitments. Despite this fact, the government was eventually forced by a combination of domestic and international pressure to affirm that a domestic emission reduction commitment was appropriate and necessary. However, it is also clear that material forces limited the domestic salience of this norm. Even after accepting that a commitment was necessary, the Thatcher and Major governments balked at the specific commitment advocated by the EU and environmental NGOs. It was not until it became apparent that emission reductions resulting from fuel switching would allow the UK to meet its commitment without additional measures that the UK was willing to bring its commitment in line with the international norm. Both Conservative and Labour governments began to justify a number of policy changes based on the need to reduce GHG emissions. Economic interests continued to structure the parameters within which these policies were pursued, but the norm achieved a high degree of political salience and is well on its way toward norm prominence. In the future, the onus may be on actors who wish to avoid their emission reduction obligations to justify their exception to the norm.

The United States presents a very interesting case study. From the beginning, the American government opposed the emission reduction commitment norm. The early framing of climate change as a scientifically uncertain, long-term potential threat minimized the urgency in the American response, and the early emphasis on the costs of reducing CO_2 emissions created a hostile environment for advocates of emission reduction commitments. The American government rejected the norm from the beginning and sought to re-shape the international normative debate to emphasize economic efficiency and global emission reductions rather than national emission reduction commitments. This was more consistent with pre-existing domestic political norms that emphasized market mechanisms and prioritized economic efficiency. Though the norm was supported by some actors within the American political system, it has yet to achieve significant domestic salience beyond the level of state and local governments (see Fogel in this volume for further discussion of this point and its implications for the future of American climate policy). Even after the Clinton administration signed the Kyoto Protocol and committed the United States to reducing its emissions by 7 percent below 1990 levels, the Senate refused to ratify the Protocol. The lack of political salience would eventually permit the second Bush administration to abandon the Kyoto Protocol. Though the administration faced domestic criticism and international condemnation, it was clear that there was insufficient political pressure to alter the administration's position.

Why did the norm not achieve the levels of salience found in the German and British cases? A strong political consensus that climate change was occurring and required a national response was critical to neutralizing opposition from industry in Germany and the United Kingdom. Governments in both countries also assured industry that its interests would be protected. The early framing of the problem in the American debate created doubts about the need to act and the potential costs and benefits of action. The early and active involvement of economic actors facing the high costs of reducing GHG emissions produced immediate obstacles to action. Their involvement was reinforced by the significant level of political influence that the energy, automobile, and other major industrial actors possessed in the American political system. Finally, the scientific pluralism of the American policy process created opportunities for opponents of action to present evidence to generate uncertainty surrounding the veracity of scientific evidence. The combination of these forces created obstacles to the norm achieving domestic salience.

The social construction of climate change as an environmental and political problem remains highly contentious and incomplete in the American context. Elections, domestic pressure, and international condemnation have not altered this situation thus far, and new presidential elections alone are unlikely to change it. The Clinton administration was unable to significantly reform American climate policy, and a post-Bush administration will face many of the same obstacles to changing policy. However, as Fogel notes in this volume, there is a great deal of activity at the local, sub-state and regional levels that is generating renewed interest in addressing climate change. International and domestic support for American emission reduction commitments continues to build, but opponents of action also continue to generate uncertainty and fear among American politicians and the broader public. Climate change remains "under construction" in the American case.

The case studies suggest that international norms can have a significant effect on national policy responses to transnational environmental problems. However, it also suggests that domestic political norms, institutional structures, and material variables influence the translation of international norms into domestic political dialogue and policy. These variables determine the international norm's level of domestic political salience. The variables shaping norm salience in the German, British, and American cases are also apparent in the Japanese and Dutch cases that are analyzed in Hattori's and Pettenger's chapters in this volume. There is a need for further research into the forces shaping the domestic salience of international norms, which may provide a firmer foundation to evaluate the interplay between international norms, domestic political salience and national behavior.

References

Abramson, R. (1989), US and Japan Block Firm Stand on Global Pollutants. *Los Angeles Times*, (November 8), p. 1.

Barnett, M. (1999), Culture, Strategy and Foreign Policy Change: Israel's Road to Oslo. *European Journal of International Relations*, 5 (1), pp. 5–36.

Boehmer-Christiansen, S. and Skea, J. (1991), *Acid Politics: Environmental and Energy Policies in Britain and Germany*. New York: Belhaven Press.

Brown, P. and Williams, N. (1990) UK: Thatcher's Global Warming Plan Criticised by Opposition MPs. *Guardian* (May 26), p.1. <www.lexisnexis.com>.

Bundesministerium fur Umwelt, Naturschutz und Reaktorsicherheit (2000), *Umweltbewusstsein in Deutschland 2000: Ergebnisse einer reprasentiven Bevolkerungsumfrage*. Berlin.

Cass, L.R. (2006), *The Failures of American and European Climate Policy: International Norms, Domestic Politics, and Unachievable Commitments*. Albany, NY: SUNY Press.

Checkel, J.T. (1999), Norms, Institutions, and National Identity in Contemporary Europe. *International Studies Quarterly*, 43 (1), pp. 83–114.

________. (1997), International Norms and Domestic Politics: Bridging the Rationalist-Constructivist Divide. *European Journal of International Relations*, 3 (4), pp. 473–495.

Cortell, A. and Davis Jr., J. (1996), How Do International Institutions Matter? The Domestic Impact of International Rules and Norms. *International Studies Quarterly*, 40, pp. 451–478.

Cortell, A.P. and Davis Jr., J. (2000), Understanding the Domestic Impact of International Norms: A Research Agenda. *International Studies Review*, 2 (1), pp. 65–87.

Department of the Environment, Transport, and the Regions (2000), *Climate Change: The UK Programme.* London: Her Majesty's Stationary Office.

ECO (1991), America at the Crossroads: A National Energy Strategy Poll. (February 19).

ENDS Report (1998), Consultation on climate Change Leaves Holes in Jigsaw. 285 (October).

Federal Republic of Germany (1994), *First Report of the Government of the Federal Republic of Germany Pursuant to the United Nations Framework Convention on Climate Change.* (September).

Finnemore, M. and Sikkink, K. (1998), International Norm Dynamics and Political Change. *International Organization*, 52 (4), pp. 887–917.

Gurowitz, A. (1999), Mobilizing International Norms: Domestic Actors, Immigrants, and the Japanese State. *World Politics*, 51 (3), pp. 413–445.

Hecht, M.T. and Tirpak, D. (1995), Framework Agreement on Climate Change: A Scientific and Policy History. *Climactic Change*, 29, pp. 371–402.

Hunt, J. (1990), Patten Proposes EC Limitations on Emissions of Carbon Dioxide. *The Financial Times*, (March 24), p. 4.

International Environment Reporter (2000), Germany Unveils Program to Meet Domestic GHG Emission Reduction Targets. 23 (22: October 25).

________. (1997), Ministers Agree to 7.5 Percent Cut in Greenhouse Gas Emissions by 2005. 20 (13: June 25).

________. (1997), Measures to Reverse Climate Change Top Agenda of IEA Ministerial Meeting. 20 (11: May 28).

________. (1996), Government Study Says Pledge to Reduce CO_2 Emissions Will Cost 270,000 Jobs. 19 (25: December 11).

________. (1989), U.K. Representative to United Nations Proposes Global Convention to Council. 12 (6: June 14).

________. (1989), House of Commons Members Attack Stance of U.K. at Noordwijk Meeting. 12 (12: December 13).

Jepperson, R., A. Wendt, and Katzenstein, P.J. (1996), "Norms, Identity, and Culture in National Security" in P. Katzenstein, ed. *The Culture of National Security: Norms and Identity in World Politics*. New York: Columbia University Press, pp. 33–75.

Kabel, M. (1990), Germany Claims World Leader Role with Anti-Greenhouse Plan. *The Reuter Library Report* (November 7). <www.lexisnexis.com>.

Klotz, A. (1995), Norms Reconstituting Interests: Global Racial Equality and US Sanctions against South Africa. *International Organization*. 49 (3), pp. 451–478.

Kowert, P. and Legro, J. (1996), "Norms, Identity, and Their Limits: A Theoretical Reprise" in P.J. Katzenstein, ed. *The Culture of National Security: Norms and Identity in World Politics*. New York: Columbia University Press, pp. 451–497.

Neale, G. and Wastell, D. (1989), Britain Fights Green Demands from EEC. *Sunday Telegraph* (November 5), p. 4. <www.lexisnexis.com>.

Payne, R.A. (2001), Persuasion, Frames, and Norm Construction. *European Journal of International Relations*, 7 (1), pp. 37–61.

Rabe, B. (2004), *Statehouse and Greenhouse: The Emerging Politics of American Climate Change Policy*. Washington, DC: Brookings Institution Press.

Rest, A. and Bleischwitz, R. (1991), "Policies and Legal Instruments in FRG to Combat Climate Change and the Greenhouse Effect" in T. Iwama, ed. *Policies and Laws on Global Warming: International and Comparative Analysis*. Tokyo: Environmental Research Center, pp. 159–181.

Risse, T. (2000), 'Let's Argue!': Communicative Action in World Politics. *International Organization*, 54 (1), pp. 1–39.

Shabecoff, P (1990), Bush Denies Putting Off Action on Averting Global Climate Shift. *New York Times*, (April 19), p.B4.

________. (1990), Bush Asks Cautious Response to Threat of Global Warming. *New York Times*, (February 6), p. 1.

________. (1989), US to Urge Joint Environmental Effort at Summit. *New York Times*, (July 6), p. 9.

________. (1989), US, in a Shift, Seeks Treaty on Global Warming. *New York Times*, (May 12), p. 1.

UNFCCC (1997), Framework Compilation of Proposals from Parties for the Elements of a Protocol or Another Legal Instrument. *FCCC/AGBM/1997/2*, (January 31).

US Congress. House. Subcommittee on Human Rights and International Organizations of the Committee on Foreign Affairs (1988), *Global Climate Changes: Greenhouse Effect*. Washington, DC: GPO.

von Moltke, K. (1991), Three Reports on German Environmental Policy. *Environment*, 33 (7: September), pp. 25–30.

The Week in Germany (1990), Toepfer: Lack of Knowledge No Excuse for Inaction. (March 20). <www.lexisnexis.com>.

Weisskopf, M. (1990), Bush Pledges Research on Global Warming; Speech to U.N.-Sponsored Panel Endorses No Proposed Remedies. *Washington Post*, (February 6), p. 1.

Chapter 3

The Netherlands' Climate Change Policy: Constructing Themselves/Constructing Climate Change

Mary E. Pettenger[1]

Climate change has a unique meaning for the Netherlands. A cursory examination of this nation, where 25 percent of the land is below sea level and 60 percent of the population lives in this area, reveals that a predicted sea level rise seriously threatens the Dutch. Intuitively, the global leadership of the Netherlands in climate change policies seems driven by self-interest stemming from its vulnerability. However, this prima facie observation is insufficient. The Dutch have continually expressed the need to promote sustainable development domestically and internationally, and contribute a vast amount of money to foreign aid, with little direct benefit perceived. For example, the Netherlands was ranked again in 2006 as the leader in "Quality" aid to the developing world (Lobe 2006). Accordingly, the Dutch are perceived, and proudly internalize the self-conception of themselves, as global leaders, agenda setters and models in the climate change arena (Pettenger 2000; Liefferink 1997; Johnson 1995). Recently, this image has been tarnished by domestic unrest (assassinations, economic stagnation and ineffective governments), limited success in reducing CO_2 emissions and the perceived failure of the Kyoto Protocol.

This chapter addresses the process of issue framing surrounding climate change in the Netherlands. The primary purpose is to identify and discuss the norms that constitute the climate change meanings held by the Dutch. This chapter asks two broad questions. What norms have influenced the Dutch definition of, and response to, climate change? Recursively, how have the Dutch redefined and promoted these norms domestically and internationally? Specifically, this study seeks to uncover the processes of norm definition, adoption, and diffusion employed by the principal norm entrepreneurs who frame the social construction of climate change in the Netherlands.

The chapter focuses on several norms identified in Dutch climate change policy that will be introduced here and discussed more explicitly in the text. The two primary

1 I wish to thank Jacco Farla and Loren Cass for their valuable comments on earlier drafts of this chapter.

norms are (1) *Sustainable Development* with several nested norms that embody the unique Dutch definition of sustainable development, and (2) *Economic Efficiency*:

- *Sustainable Development*,
  - Stewardship,
  - Equity,
  - Polluter Pays Principle,
  - Precautionary Principle, and
- *Economic Efficiency*.

Sustainable Development is defined by the Dutch as economic development that places equal emphasis on the economy and the environment, and as requiring economic, political, and social processes to be changed today to avoid passing pollution problems to future generations. Several nested norms encompass the subtle meanings within this dominant norm. First, the Dutch cultivate a *Stewardship* norm in their perceived role of caretakers of the environment for future generations. This norm arises from the Dutch past and particularly their interactions with the North Sea. Second, the *Equity* norm pervades Dutch rhetoric and policy in the promotion of equitable economic growth for developed and developing states. In essence, the Dutch leadership role as aid donors to third world states extends to their domestic and international environmental policy as well. Third, the Dutch have fully adopted the norm of *Polluter Pays*, arguing in domestic and international settings that those responsible for greenhouse gas (GHG) emissions must be responsible to eliminate or control them. The Dutch accept the argument that developed states must take the leading role in reducing emissions and that they must make domestic emission cuts. Fourth, the Dutch also have instantiated the *Precautionary Principle* norm, arguing that actions are required today to have "no regrets" in the future. In contrast and often in conflict with the norm of *Sustainable Development*, the Dutch promote the norm of *Economic Efficiency*. They have at times sought the most cost-effective, economically-viable approach to GHG emissions (following closely the discourse of ecological modernization presented by Bäckstrand and Lövbrand in this volume). This norm also follows the long Dutch tradition as merchants.

The chapter is divided into two sections. The first section presents the material and social contexts within which Dutch climate change norms are embedded. The second section illustrates the connections between the material and social contexts, and the social construction of Dutch climate change policy over the last two decades within the context of the process of climate change norm adoption, diffusion, and instantiation.

Constituting Climate Change

Understanding the constitution of climate change and the subsequent actions by the Dutch regarding climate change requires an awareness of the social and material contexts of the Netherlands. The Dutch understanding of climate change is molded from several perspectives:

- Threat of the Sea,
- Relationship with Nature and Physical Space,
- Individual versus Collective Needs, and
- Calvinist and Merchant Traditions.

These factors unite to form a particular meaning about climate change held by the Dutch, creating an internalized perspective toward nature and a driving requirement for planning and action.

The Netherlands is extremely vulnerable to climate change and environmental degradation as it is buffeted by water on the east, north, and west. Some argue that the fundamental reason for Dutch action against climate change is the extreme threat sea level rise poses to their survival. For example, "if other nations do not follow its example, the Netherlands is doomed. That is why it is so heavily involved in environmental education and negotiation at the diplomatic level" (Johnson 1995, 56). The Netherlands has been fighting water for centuries at great cost both monetarily and in loss of human life. The 1953 flood caused by high tides and strong winds killed over 1,800 people. As the estuary for three major rivers coming from the East, the heavily populated, low-lying areas of the Netherlands are always subject to flooding. This threat was most recently illustrated by the displacement of over 300,000 persons in the Rhine flood of 1993. The life and death struggle of the Dutch with water is not simply a historical item of interest but an ongoing struggle which survives into modern times, causing multiple problems in everyday life.

Besides life-threatening consequences, rising sea levels may lead to loss of wetlands, erosion of beaches and dunes, and increased salinization of groundwater and farmland due to the influx of salt water. Important infrastructure such as sewer systems, locks and the port in Rotterdam could be seriously affected by the potential necessity of pumping out rising water. It is projected that by 2050 it will cost the Netherlands "US $12.5 million a year" to adapt to sea level rise (Hoes, Schuurman and Strijker 2005, 44). Additionally, sea level rise would be exacerbated by pollution brought in from the Rhine. At this time "much of the Rhine water ... is often so polluted that it has to be flushed to the North Sea as rapidly as possible, and trapped as little as absolutely necessary" (Hekstra 1986, 320). The negative impact of sea level rise on the Dutch cannot be ignored, nevertheless, the Dutch have the experience, resources and technology to adapt well to these potentially drastic changes. In addition, one could argue that they are uniquely prepared mentally as they have battled the sea for centuries.

Some studies suggest that the effects of climate change also could be positive for the Dutch. The Netherlands Environment Assessment Agency (NEAA) reports, in addition to the potential negative consequences, that warmer temperatures may lead to an "extension [of the] Dutch tourist season" as southern European states become too hot, and a "longer season" for agriculture leading to "higher yield ...

[and] chances for other crops" (NEAA 2005, 13). Energy costs may be reduced because of an increase in mean temperature and less energy needed to heat buildings. Air conditioning costs may rise but this would depend on the overall temperature increase and the type of buildings (VROM 1994a, 69).

Dutch climate change policies are embedded in the confluence of their material and social realities. Danger posed from the sea is a material reality that has driven the actions of the Dutch for centuries. However, this reality is framed as well by the social reality encapsulated in the famous saying "God made the world, the Dutch made the Netherlands." Their successful reclamation of land wrested from the North Sea has had a profound influence on their beliefs and practices. In short, "the polder—a strip of land redeemed from the sea—is a symbol of the Dutch collective consciousness for the successful struggle against threatening inundation" (Germino 2001, 783). The Dutch manufactured the polder system through a series of dikes, pumps, and waterways to reclaim land for human use. Thus, the territorial creation of the Netherlands and the relationship with the sea has created a particular identity and understanding of their surroundings. The image of "watery land" has permeated the Dutch and European psyche since the time of the Dutch prominence as seafarers and the many conquests for control of Europe. As noted by Schama (1987, 263), Napoleon ridiculed the Dutch, much to their distaste, calling their land the "'alluvium deposited by some of the principal rivers of my empire.'" The Dutch perceive themselves as part of their environment, holding a symbiotic relation with the land, and yet in control of nature.[2] Thus, their approach to climate change is more than just a response to the threat of sea level rise. It is the continuation of the struggle to create and protect themselves in the planning of their physical space. Consequently, the Dutch have come to strongly perceive nature as something to be managed, not set free, and define their roles as "stewards" of nature (Farla 2006).

The relationship to nature and physical space becomes more complex when combined with the apparent paradox between strong individualism and collective action by the Dutch. Traditionally, the Dutch have addressed environmental problems on a local level, resisting the intervention of the national government; however, the national government has assumed a greater role in driving environmental policy (Liefferink 1997). Yet, there has always been a coexistence of local and centralized response to nature. Monumental efforts to battle the sea over centuries have left a legacy of collective action necessary for the preservation of life, including highly powerful Water Boards that manage the vast polder systems.

The understanding of the Dutch identity and relation with land must also include a brief discussion of the Calvinist influence with its consociate work ethic. While the material reality of the sea necessitated efforts to reduce vulnerability, the spiritual drive ordained by Calvinism to demonstrate the worthiness of one's soul is also

2 This physical and social manifestation of territory provides an interesting corollary to the concept of territoriality as discussed by Paterson and Stripple in this volume. In a time when sovereignty is defined by land, the Dutch are in truth, constantly constructing themselves.

of extreme importance (Rochon 1999, 26–29). The need to demonstrate Christian virtue is visible in the mythical Drowning Cell of the 1600s. Intractable prisoners were placed in a hole below sea level with a pump to work or drown, serving as a lesson to cure the "lazy" prisoners by teaching the work ethic. The practice assumes deeper significance in the Dutch context as the struggle against the sea is a matter of life and death. Operating as an important national myth, the rumors of the Drowning Cell's existence feeds a population who is continually reminded of the need for constant vigilance, and who must prove worthiness by perpetually holding back the sea (Schama 1987, 22–25).

Finally, Dutch identity is framed by its history. During the seventeenth century, the Dutch dominated the seas, establishing colonies and strong international trade. The "merchant" tradition (Farla 2006; Hendriks and Toonen 2001, 4) continues today with strong support for commercial interests, and a historical memory and pride that this small state was once a great sea power (also defined as "maritime commercialism" by Voorhoeve 1979, 26). Likewise, the Netherlands is a highly industrialized state that is heavily dependent on exports and imports, and commands respect in the international community.

Consequently, social and material influences combine to create a unique perspective on, and impetus for, Dutch climate change policy. The Dutch can be characterized as having a strong sense of proactive and proud stewardship of their land, coupled with a passionate and pragmatic approach toward solving the climate change problem. These characteristics as well shape the *Sustainable Development* and *Economic Efficiency* norms they have adopted. On the one hand, they seek to preserve themselves and their land through sustainable practices, while on the other they seek to promote their economic well-being. Informed by the unique social and material circumstances of the Dutch, the players involved in environmental policy in the Netherlands and the particular policies created by the Dutch will be outlined.

Domestic Context: The Netherlands' Environmental and Climate Policy

Dutch environmental consciousness and policymaking was significantly altered in the late 1980s by the impact of two publications. Significantly, the first document introduced the norm of *Sustainable Development* to the Netherlands emerging from the *Brundtland Report* issued by the United Nations (WCED 1987). Domestically, in 1989 the National Institute of Public Health and Environmental Protection (Rijksinstituut voor Volksgezondheid en Milieuhygiene—RIVM) produced the second document, *Concern for Tomorrow* (*Zorgen over Morgen*). The survey of the Dutch environment predicted that environmental problems would worsen in spite of efforts to clean up their environment and reduce pollution. These two documents had a significant impact, along with severe weather in the late 1980s, as the threat of environmental degradation to Dutch quality of life became extremely salient. The response in 1989 was the creation of a new approach to environmental policymaking with the first National Environmental Policy Plan (NEPP1).

Domestic Actors

Political parties and their subsequent ruling coalitions through the Dutch government ministries have been the primary actors in the Netherlands climate change policy promotion. The primary ministry responsible for environmental policymaking and implementation is the Ministry of Housing, Spatial Planning and Environment (Ministerie van Volkshuisvesting, Ruimtelijke Ordening en Milieubeheer—VROM) established in 1982. The NEPP1 was the result of joint policymaking between VROM, the Ministry of Economic Affairs (Ministrie van Economische Zaken—MEZ), the Ministry of Agriculture and Fisheries, and the Ministry of Transport and Public Works. Significantly, NEPP1 was produced by the centrist-right Christian Democratic Appeal Party (CDA) which was re-elected based on its support for the document. Gradually, the number of ministries involved in environmental planning have expanded from the initial four to include: the Ministry of the Interior and Kingdom Relations; the Ministry of Foreign Affairs (Ministrie van Buitenlandse Zaken—BuZa), the Ministry of Finance; the Ministry of Health, Welfare, and Sports, and the Minister of Education, Culture, and Science. However, the increase in ministries should not be construed as a sign of greater importance of the issue, but rather as a process of policymaking in which more stakeholders are claiming a portion of the issue.

Domestically, VROM must work closely with the Ministry of Foreign Affairs as Dutch environmental policy has focused increasingly on the international level. BuZa plays a large role in pushing Dutch policy positions in the international arena, e.g., in international organizations such as the European Union and the United Nations, and unilaterally with individual states through embassies and diverse climate change projects. Consequently, BuZa participates in the domestic creation of Dutch environmental policy by representing the Netherlands' international interests.

Other ministries, such as the Ministry of Agriculture, Nature Conservation and Fisheries, the Ministry of Transport and Public Works, and the Ministry of Economic Affairs, also play a role in environmental policy formation. These ministries control areas that are integral to environmental policy such as water quality or agricultural waste. However, they often share policy arenas with conflicting interests and goals, such as in the case of public transportation systems or that of the expansion of Schipol airport (van der Straaten and Ugelow 1994, 118). Nevertheless, joint policymaking between ministries has become more common in recognition of the need to coordinate environmental planning in overlapping policy arenas.[3]

3 This study purposely leaves out interest groups in Dutch environmental policy making. The Dutch citizenry generally are supportive of environmental causes. Two prominent environmental groups are headquartered in the Netherlands, Greenpeace and Friends of the Earth (as noted by Rochon (1999, 53) in the 1990s "over 20 percent of the worldwide membership of Greenpeace was Dutch"). However, political parties often adopt pro-environmental positions and the government has been pro-active on environmental causes, thus reducing the need for heavily active environmental interest groups (Cramer 1989; Dijkink and van der Westen 1992; Liefferink 1999b).

Domestic Policies

This chapter highlights two types of documents, National Environmental Policy Plans (NEPP) and Climate Memorandums/Notes, to compare and contrast the Netherlands' climate change policy from 1988 to 2006. The Dutch produce highly innovative planning documents that promote strong actions to address climate change; however, it should be noted that often the rhetoric has not always been matched to concrete policies and implementation. These primary data sources document the interplay of social and material contexts, actors, and processes of issue framing, through norm adoption and diffusion. The documents will be briefly introduced chronologically to demonstrate the processes of change within the social construction of climate change in the Netherlands.

Climate Change Meanings: Context, Norms and Actions

Initially, the impetus to address environmental degradation emerged from a top-down perspective in the Netherlands, with the government playing a leading role in defining the issues and their importance. 1989 has been called the high point of environmental consciousness in the Netherlands (Liefferink 1997; Johnson 1995). Queen Beatrix, in her Christmas message of 1989, stressed the growing threat of pollution and the need for a Dutch response:

> The earth is slowly dying.... We human beings ourselves have become a threat to the planet. Those who no longer wish to disregard the insidious pollution and depredation of the environment are driven to despair (as quoted in *Economist* 1989a, 45).

This time of heightened environmental awareness led to a significant change in Dutch environmental policy.

National Environmental Policy Plan (NEPP1)

On May 25, 1989, the Tweede Kamer (the Lower House of the Dutch Parliament) was presented with the National Environmental Policy Plan (NEPP1) cogently titled *Choose or Lose*" (*Kiezen of Verliezen*) (VROM 1989). Representing the first defining example of norm contestation and adoption in Dutch politics, its publication led to an amazing event in Dutch politics—the Dutch government fell over an environmental policy. The coalition government of the CDA and Liberals (VVD) fought over financing of the NEPP. The CDA wanted to obtain funds from taxes on car owners, gasoline taxes, and the private sector. The Liberals charged that the CDA plan would have placed an unfair tax burden on car owners, arguing that the NEPP should be funded with profits from economic growth (*Economist* 1989a, 45). Prime Minister Lubbers stepped down, leaving it to the voters to determine Dutch support for the new environmental plans, and ending the center-right coalition of the CDA and VVD who had held majority power from 1982–1989.

What followed was "the greenest election campaign the world has ever seen" (*Economist* 1989b, 22). In September 1989, the CDA was overwhelmingly supported by "35% of the vote," the VVD lost seats, and a new right-left government was formed with the CDA and Labor Party (Partij van de Arbeid—PvdA) (*New Scientist* 1989, 24). As important as this election was domestically, the debate which prompted the election would prove to be globally significant.

Internationally, the NEPP1 was one of the first official documents to specifically address climate change as a serious environmental problem (Liefferink 1997; Metz and Vellinga 1990). Additionally, the NEPP "came to be seen as one of the first and most eloquent expressions of a new way of thinking in environmental policy" (Liefferink 1999a, 256). Importantly, the NEPP has served as a "source of inspiration" for other states to study and in some cases emulate (ibid., 274). For example, the Fifth Environmental Action Programme *Towards Sustainability* (European Community 1993) published by the European Union (EU) in 1993 was "partially inspired by the Dutch approach" found in the NEPP (VROM 1999a, 25). The similarities include, the "overall structure ... [the] importance of clear information on policymaking ... [how it]designates five target sectors, analogous to the target groups of the NEPP ... [and] recognizes environmental effects through the designation of themes" (de Jongh and Captain 1999, 208). In addition, the NEPP has been "propagated to a large number of individual countries" such as China, Latvia, Hungary, Austria, and to the US states of New Jersey and North Dakota (Liefferink 1999a, 273). Domestically, the NEPP assumed critical importance and framed the conception of the norm of *Sustainable Development*.

The NEPP1 is an ambitious, comprehensive plan designed to guide future policymaking.[4] Since the adoption of the NEPP1, the national plans have served as guides for all subsequent governmental policies and laws, and precipitated five major changes in political perspective. The first major shift in Dutch environmental policy was a unification of policy around eight environmental themes. The Dutch reoriented their attack on environmental degradation from a sectoral to a comprehensive approach, viewing the environment as an organic whole with a need for long-term planning. This shift also influences the norm of *Stewardship* as the Dutch view their domestic environment as part of the global environment.

The second conceptual shift in Dutch environmental policy was the division of Dutch society into "target groups," for example, agriculture, industry or consumers who contribute to pollution and must be part of the solution to limit emissions. This signifies a further incorporation of the Dutch definition of *Sustainable Development* and *Polluter Pays Principle* by emphasizing the relationship between human action and the environment. All parts of society contribute to pollution and all must act responsibly to reduce and eliminate pollution.

---

4 NEPP2 "The Environment: Today's Touchstone" was published in 1994 and NEPP3 was published in 1998; however, due to the length of this chapter they will not be discussed. Many of the principles established in NEPP1 were upheld and strengthened.

The third shift in policymaking is demonstrated by the amount of consensus present from the planning stages of the NEPP, to the agreements between the target groups and the government on how to achieve the NEPP's environmental goals. The government involves target groups in the policymaking process as one means of generating public support for its environmental policy in a two-pronged approach with incentives and disincentives. First, in the policymaking stage, the government must pay "serious attention to the needs and wishes of groups affected by environmental policies and taking them into account in the design of measures" (Liefferink 1997, 229). This means the government learns more about a particular sector, and when the government makes the final decision, hopefully it will be made with the buy-in of the sector. Consequently, all of the NEPPs embody the ever-evolving consensus relationship (Andeweg and Irwin 2002, 209–213; Hendriks and Toonen 2001) that exists between the Dutch government and Dutch society (Liefferink 1999a, 266; Tuininga 1994, 12).

This somewhat successful strategy has not been easy or perfect. For example, the metals industry was reluctant to make the heavy cuts necessary. In 1992 the Environment Minister Hans Alders "issued an ultimatum to the metals industry: if it does not agree by 20 January [1992] to curb pollution it will face stringent legislation," thus echoing the national government's stand that those industries that do not "voluntarily toe the environmental line" will "face stringent legislation" (*New Scientist* 1992, 11). Additionally, early in the process, the chemical industry was hesitant in some cases to accept the goals set out in the NEPP. In particular, they disliked a drive to reduce CO_2 emissions by imposing a tax on emitters: "Industry, particularly the chemical industry, has reacted in outrage ..." declaring that "imposing a carbon tax without other European countries doing the same would be suicide" (Bierdermann 1992, 164). Importantly, this position reflects the merchant identity of the Dutch based on the norm of *Economic Efficiency*. Costly changes should not be taken unless the burden can be shared by many.

Consequently, the NEPPs embody the centralized, consensus-based relationship that exists between the Dutch government and society. The government plays the dominate role in norm definition, stipulating the objectives to be reached in emission reduction, and working closely with the target group to determine how to reach the goals.

The fourth shift brought forward by NEPP1, embodies the principle of *Equity* nested within the Dutch norm of *Sustainable Development*. The NEPP1 stipulates that changes must be made today to avoid passing pollution problems to future generations:

> With rights go responsibilities: the responsibility of every individual to manage the environment, that is, to carry out those activities or institute those measures which are necessary for the attainment and preservation of adequate environmental quality. Adequate not only for the current generation but also for future generations (VROM 1989, 39).[5]

---

5 In my study of Dutch statements in the United Nations, the idea of environmental well-being alongside ecological development was found as early as 1969 (United Nations

The fifth significant policy shift is the importance placed on international environmental policy, called a "global view." This places the Dutch squarely in the dominate discourse that defines climate change as a global issue that requires a global response as defined in several other chapters in this book. Due to the international nature of many environmental problems faced by the Dutch, they perceive the need to develop a comprehensive domestic and international approach. The Dutch act most often within the context of the European Union to promote their climate policies. However, as noted in the beginning, they are also leaders in development aid, and strongly believe that international responsibility is a necessity to fulfill their identity as wise, responsible stewards, grounded in the norms of *Stewardship* and *Equity*. I turn now to outlining the formation of Dutch climate policy.

Netherlands' Climate Change Publications

Climate change became a salient issue in the late 1980s in the Netherlands primarily because of growing scientific information. The emergence of the *Precautionary Principle* norm is prevalent in Dutch documents from the inception of the NEPPs and the climate memorandums. The Dutch quickly accepted the scientific evidence that climate change was a reality that would threaten the Netherlands. In response to this perceived threat, they devised numerous strategies to reduce their, and other states', GHG emissions, emphasizing a multilateral, state-focused approach to the problem (similar to the pattern outlined in the Cass and Hattori chapters in this book). In addition, as Dutch environmental consciousness rose, so did self-awareness of their contribution to this global problem. The Dutch have adopted the norm of *Equity* to mean they are responsible for their own behavior, and must act to protect themselves and others. However, this norm is challenged by two important material realities. The Dutch produce a significant portion of per capita emissions of GHG in comparison to other states in the European Union. Likewise, they must face the inescapable reality that the Netherlands is an energy-intensive, developed society which tends to favor a cost efficiency approach to solving problems. In this case, their material reality combines to support the *Economic Efficiency* norm, standing in contrast to the norm of *Equity*.

In response to conditions set out in the United Nations Framework Convention on Climate Change (UNFCCC) in 1994, the Dutch government published the *Netherlands' National Communication on Climate Change Policies* (VROM 1994b). The *Communication* outlines the particular characteristics of environmental problems in the Netherlands, provides an in-depth inventory of emissions, presents domestic strategies to reduce emissions, and assesses the results. The document also advocates

---

1969, 205–6), but not conceptualized as sustainable development until after the publication of the Brundtland Report.

the particular international approaches the Dutch favor such as joint implementation and international cooperation to finance sustainable projects in developing states.[6]

Subsequent documents have altered these goals because CO_2 emissions by the Dutch (and the world) have continued to escalate since the publication of NEPP1 in 1989. For example, the Netherlands saw its level of CO_2 contribution increase by the year 2000 rather than decrease since 1990 (from 210.3 million tonnes of CO_2 in 1990 to 216.9 in 2000). The top two contributors, Germany and the United Kingdom, saw their overall contributions decrease (Germany 1222.8 million tonnes in 1990 to 991.4 in 2000, and the United Kingdom 742.5 million tonnes in 1990 to 649.1 in 2000) (European Environment Agency 2003).

Consequently, the Netherlands revised its goals and policies to aim for a more realistic GHG emission reduction timetable. The Dutch joined with the rest of the European Union to sign the Kyoto Protocol in 1998, and in conjunction with requirements in the treaty they produced two documents. In June 1999, Part I of the *Climate Policy Implementation Plan* (VROM 1999b) devoted to domestic measures was published and, in March 2000, Part II (VROM 2000) devoted to international measures was released. Significantly, Dutch social construction of climate change is evident in the adoption of the norm of *Equity* in the conscious decision to require domestic reductions, while at the same time promoting the norm of *Economic Efficiency* by supporting more cost-effective international reductions. Thus, and in collaboration with the EU, the Netherlands believes that 50 percent of its reductions should come from domestic sources and 50 percent from international efforts. In sum, Part I provided three approaches to decrease emission based on the most cost-effective means, yet responsive to progress plans. In addition, it stressed the need for technological and instrumental innovation (VROM 1999b). Dutch international policy was refined in Part II to include the measures prescribed in the Kyoto Protocol (Clean Development Mechanisms, Joint Implementation and Emission Trading). The stated goal of the documents is the EU standard of a 6 percent reduction in GHG emissions to 1990 levels by 2008–2012.

The Dutch government promoted these norms globally by hosting the November 2000 Sixth Conference of the Parties (COP6) held in the Hague, chaired by Minister Pronk of the Dutch Environmental Ministry. The Dutch government's and the citizens' optimism about a multilateral approach to climate change was severely wounded when the meeting failed to come to agreement on measures to meet the Kyoto Protocol goals. A new round of meetings was called for in June 2001 by Minister Pronk, and held in Bonn Germany (COP6-II). However, this meeting also failed to overcome major obstacles to gaining US support for the Kyoto Protocol.

6 Follow-up publications include *The Second Netherlands' Memorandum on Climate Change* (VROM 1996) and the "Update" (VROM 1998a) to this document.

Netherlands Environmental Policy Plan 4

In 2001, the NEPP4, *One World and One Will: Working on Sustainable Development* (*Een wereld, en een wil: werken ann duurzaamheid*) was published (VROM 2001). The NEPP4 has presented a shift in environmental focus by following the theme *A New Look: Broader and More Future-oriented Vision* (*Een Nieuwe Kijk: Breeder and Verder Vooruit*). Continuing the global focus of the Dutch rather than focusing more heavily within the Netherlands, the document sets out seven environmental problems that all humans face over the time period to 2030: Loss of Biodiversity, Climate Change, Over-exploitation of Natural Resources, Threats to Health, Threats to External Safety, Damage to the Quality of the Living Environment, and Possible Unmanageable Risks. The shift to broader themes emphasized seeking common solutions in an ever-increasingly globalized world, looking further into the future to be able to "make [wise] choices today" (ibid.). Clearly, this document establishes the credibility of the *Precautionary Principle* in Dutch policy while continuing to frame climate change as a global issue.

The NEPP4 articulates even more directly the tenet of *Sustainable Development*, from the title, "A New Look: Broader and More Future-Oriented," to its refocus on issues of global concern. The following excerpt from NEPP4 is a clear example of the language of sustainable development in the Dutch context closely resembling statements from NEPP1 in 1989:

> The public authorities in the Netherlands are charged with enhancing the well-being and living standards of all its inhabitants both now and in the future... *enshrined in Article 21 of the Constitution.* Sustainable development requires not only that the environment is clean and free of pollution and nuisance, but also that good-quality natural resources are available for all, both now and in the future. *It is vital that the distribution of and access to natural resources are fair, not only within the Netherlands but also globally.* [italics mine].... sustainable development can only be achieved in the Netherlands in an international context, recognising that the Netherlands forms part of a larger whole in social, economic and ecological terms (VROM 2001, 11–12).

At least from the government's perspective, it appears that maintaining economic growth at the expense of environmental quality is perceived as wrong and has been more readily accepted into Dutch self-conception at the ministerial level. However, one cannot ignore that the diffusion of the norm of *Sustainable Development* in the general population has generated tensions since its introduction. Economic problems, such as unemployment, have often dominated Dutch political actions. In addition, the Dutch strongly value their merchant identity in the world economic market, gain a great deal of revenue from their natural gas fields, and are economically dependent on international trade.

The NEPP4 also sought to set strikingly high targets for emission reductions, including a "40 to 60% emission reduction as compared to 1990" as one of the "normative objectives for Western Europe for the year 2030" (ibid., 33). While the Dutch government established an energy tax in 1996, the NEPP4 clearly shows the Dutch conviction in a global and EU-based response to address emission reductions:

> the energy issue is a global problem.... As long as global co-operation makes progress in the area of CO_2 reduction, the Netherlands will do its utmost to make further agreements and to comply with them. If this co-operation falters, the Netherlands will continue to do its utmost to reduce other emissions (in the EU context) (ibid., 33).

However, NEPP4 was born during a time of extreme transition in Dutch politics. Efforts to reduce GHG emissions seemingly failed as emissions rose rather than declined, public money was being spent to enlarge rather than decrease polluting industries such as the expansion of the national airport (Schipol), and strong political pressure rose surrounding the issues of unemployment and economic recession. Additionally, the NEPPs had proven successful only to a point, "considering the difficulties encountered in implementing the NEPP.... Even with internalization, in the end, turkeys are not likely to vote for Christmas" (Liefferink 1999a, 272).

Consequently, CO_2 emissions continued to rise, and the Netherlands was unable to meet its stated goal of a 3 percent reduction to 1990 levels by 2000. At the same time, the Dutch psyche was also deeply damaged by the failures of the COP6 in the Hague, and the subsequent United States withdrawal and rejection of the Kyoto Protocol. The NEPP4 was also authored by the Labor Party (PvdA) government of Wim Kok which had proudly promoted Dutch environmental policy as an example for the world to follow. The future was to change quickly by the time the document was published.

Recent events have challenged Dutch abilities and reputation, and demonstrate the struggle over norm adoption and diffusion in the Dutch context. First, the elections in 2002 and 2003 disrupted the smooth flow of politics in this state that has been previously described as politically stable, a definition now at odds with the pre-election assassination of Pim Fortuyn, leader of the extreme right-leaning List Pim Fortuyn party (LPF). In the subsequent election after his death, his new party received 26 percent of the vote, with the center-right party (CDA), receiving 43 percent and returning to power for the first time since 1994. The subsequent attempt to form a coalition government with the LPF and CDA led to political stagnation and the eventual fall of the government only six months later. New elections resulted in the CDA receiving a plurality of 44 percent and now leading a coalition government (Election World 2004). Second, the gruesome assassination of Theo van Gogh in November 2004 by an Islamic extremist has shifted the focus of the already nervous Dutch more forcefully to the issues of terrorism and immigration.

Coupled with these national concerns, the role of the economy has a significant impact on Dutch climate change policies. The Dutch economy was negatively affected by the economic turmoil following the events of September 11th. Ironically, even with the economic downturn, CO_2 emissions continued to climb and by 2001 had grown 4 percent since 1990 (VROM 2002, 19). Today, however, their Gross Domestic Product has made a strong rebound from $398.5 billion in 1999 to $621.3 billion (estimated) in 2005 (OECD 2006), thus increasingly challenging their ability to cut energy use and emissions.

More recent events in the early 2000s seem to demonstrate further erosion of the norm of *Sustainable Development* and the *Polluter Pays Principle* within the Dutch context. The confluence of domestic and economic turmoil have reduced the trust the Dutch have in their government, rising to the point where the citizens of this small state known as a staunch EU supporter overwhelmingly voted against the European Union constitution counter to the advice of the government and the major political parties. Concurrent to this loss of faith in the government, the Dutch population has lost interest in the environment, as new issues such as unemployment and terrorism are deemed more significant. Clearly, the election of a conservative government in the Netherlands in May 2002 initially led to a decrease in progressive environmental policy. As the State Secretary for the Environment, Pieter van Geel said "It is nostalgia to profile The Netherlands as the leader it might once have been in the area of environmental issues" (Shared Spaces 2003). In addition, the temporal nature of climate change pushed the issue out of Dutch minds. Climate change was perceived as "a less immediate issue. People are bound to regard a problem they are not likely to notice for some years as less urgent" (Verhoog 1999, 3).

Dutch and European Union Environmental Policy More recently, however, the Dutch are reinventing their climate policy and attempting to resume their role as environmental leaders in Europe.[7] The Dutch have placed great importance on the European Union as a vehicle to promote their environmental policies from early on (Liefferink 1997; Tuininga 1994; van der Straaten and Ugelow 1994; de Jongh 1989). The Dutch norm of *Sustainable Development* with the nested norms of *Stewardship*, *Equity*, *Polluter Pays* and the *Precautionary Principle*, alongside the competing norm of *Economic Efficiency* are all present in the EU's policies, making it hard to determine the flow of influence between Dutch domestic and EU organizational norm adoption and definition. Being a small state with a small voice, the Netherlands "has sought various strategies to 'use' the EU to push the global climate change policy process" (Andersson and Mol 2002, 60).[8]

As a member of the EU, the Netherlands is held to support the position of the organization, even if it desires to negotiate on its own with states outside the EU. For example, the Dutch strongly favor the implementation of the Kyoto Protocol

7 The Dutch government also published the *Action Programme for Sustainable Development* in 2003 summarized in *Sustainable Action*. This document examines and guides the implementation of sustainable practices in all sectors and levels of the government to be coordinated by VROM domestically, and in a separate document "international strategy" is to be coordinated by BuZa (VROM 2004a and 2003).

8 The interaction of Dutch and EU environmental interests is complex. Clearly, unified policy in the context of the EU can strengthen the Dutch position (Liefferink and Anderson 1998, 267–8). At times however Dutch environmental policy has been more stringent that the EU and the Dutch have to accept the loss of sovereignty (Liefferink 1996, 15; McCormick 1999, 195–202).

and the ratification of the treaty by the United States. However, they are unable to alter the positions of other EU states that will not change their policies to fit the US agenda: "Thus, the Netherlands may have a dilemma in how to deal with a situation where it has to broke[r] deals both at the EU level and global level" (ibid., 62). Additionally, the EU decision making process is "slow, inflexible and introvert" (ibid., 64) making it difficult for the Netherlands to use its motivation and expertise to bring the dramatic change it desires.[9] Finally, owing to the fact that the EU is primarily at this time an economic international organization, independent actions by the Dutch can serve to damage their trade relations with the other members, i.e., creating unfair trade advantage, or loss of competitive advantage (de Jongh and Captain 1999, 257). The nature of the European Union plays a large role in shaping Dutch norms. Nevertheless, the Dutch rarely go it alone in EU policy (Liefferink 1997, 245).

One means of uncovering the influence of Dutch policy on the European Union is found in the topics they promote during their EU Council presidency. The most recent Dutch presidency from July 1 to December 31, 2004 demonstrates the reemergence of the Netherlands as a climate change leader in the EU. For example, while serving as EU Council President, the Dutch promoted their "Clean, Clever, Competitive" approach (VROM 2004b) as part of the Lisbon Strategy to create an EU that is "the most dynamic and competitive knowledge economy in the world by 2010" while "reinforcing the 'environmental pillar' of the strategy" (EU 2004). This theme was taken up again after the Dutch left the presidential seat in March 2005 when the EU environmental ministers met to discuss the Kok Report and to "propose that developed countries should consider cutting their greenhouse gas emissions by 15–30% in 2020 and 60–80% in 2050" led by Dutch environmental minister Van Geel (VROM 2005b). *Economic Efficiency* is heavily promoted in this approach as the Dutch government states "the Kok Report assumes the position that environmental innovation is good for the economy and employment … . [and views] the environment as an opportunity to improve European competitiveness" (VROM 2004c).

In addition, the Dutch prime minister Jan Peter Balkenende and the UK's prime minister Tony Blair, called for climate change to be "the top of the EU's agenda" at the EU Council meeting in October 2006. They joined forces to pressure the EU to take immediate action (supporting the *Precautionary Principle*), "We have a window of only 10–15 years to take the steps we need to avoid crossing a catastrophic tipping point" (Laitner and Parker 2006, 8).

Recent Domestic Policies Alongside the EU, the Dutch government is working to reduce transportation emissions by promoting alternatives to fossil fuel, including renewable energy sources and biofuels. The Dutch government is requiring "By 2007,

9 Even this small, pro-environmental state has its own environmental problems and has tried to avoid imposition of stronger legislation coming down from the EU level. In one case, the Dutch resisted EU legislation to reduce eutrophication (Liefferink 1997, 237).

2% of the petrol and diesel brought onto the Dutch market must contain biofuels. ... to be increased to 5.75% by 2010." Significantly for the norm of *Sustainable Development*, these biofuels must "meet minimum sustainability requirements" by banning products from "large-scale deforestation" (VROM 2006c).

Addressing energy use is also high on the climate agenda. The first green tax on energy was introduced in the Netherlands in 1996 and was "aimed at reducing carbon dioxide emissions from small users while maintaining the competitive position of the large industrial users" (Zito et al 2003, 167). Funds collected are funneled back into the economy through financial incentives for increased energy efficiency and renewable energy. This tax remains in force, slowly increasing over time. Additionally, the focus on energy consumption has been more recently addressed internationally as well as seen in the Netherlands' hosting of the 2004 World Conference "Energy for Development." As noted in the NEPP4, sustainable energy consumption is vital for both the developed and developing world. Enforcing the *Equity* norm, the Netherlands projects itself as a leader in guiding the developing world to sustainable energy practices (albeit as it defines them). In addition, in October 2006, the Dutch government hosted "Make Markets Work for Climate" to discuss "climate friendly investment in the energy sector." Participants included the Dutch prime minister, president of the International Energy Agency, Shell Oil, ABN-AMRO (the largest bank in the Netherlands) and the World Bank (VROM 2006b).

The Dutch have taken other significant actions on climate change, many reflecting a growing tendency to adopt the *Economic Efficiency* norm. While the Dutch were part of the European Union's initial critique of the inclusion of emission trading as a Kyoto mechanism, they have now fully adopted this measure and are proud of their newly created emission trading markets. In addition, the Dutch were one of the first European states to set up two emission trading systems for NO_x and CO_2 emissions in 2005 (VROM 2005c).

Concurrently, one highlight of Dutch actions on climate change is their leadership in the Clean Development Mechanism (CDM) and promotion of the *Polluter Pays* and *Equity* norms (as well as *Economic Efficiency* in seeking international, cost-effective solutions). In January 2002, they were the first state to sign an "international contract with the World Bank to develop clean energy projects in developing countries to help slow global warming" (Reuters 2002). This number has grown to 18 CDM projects in 16 states, including a June 2006 agreement with China. The Netherlands has been assigned "100 millions tons" of international CO_2 emission reduction of which their "CDM projects are intended to achieve a reduction of 67 million tons of CO_2" (VROM 2006d).

The future of climate change policy appears promising again for the Dutch. The Economic Ministry 2005 Energy Report noted pride in the Netherlands' industry ranking as the most energy-efficient in the world (MEZ 2005, 20) while recognizing that emission growth will continue to climb globally, requiring Dutch attention. Additionally, Dutch businesses who received emission permits "emitted 7% less CO2 than they were allocated emission rights [and] 18% less NOx" (VROM 2006a).

Climate policy is back on the Dutch agenda with the Climate Policy Evaluation Memorandum 2005: *On the Way to Kyoto* (VROM 2005a). The publication gives a comprehensive overview of Dutch climate policy to date with several items of pride for the Dutch. It fulfills the motto on all Ministry of VROM documents: "Knowing that, in a small country like the Netherlands, it pays to think big." VROM states that it can "deliver [with] 90% certainty that the domestic target of 220 Mtonnes will not be exceeded during the Kyoto period" (ibid., 10). The ministry is focusing on "reserve measures" to insure that the Netherlands complies with its Kyoto requirements, including "increasing the regulatory tax on natural gas ... capturing and storing CO2" and promoting biofuels. In addition, it recognizes that while it was a frontrunner in adopting JI and CDM, this market faces uncertainty in terms of implementation and design, and may face increased competition as well as higher prices (ibid., 13–14).

The Dutch also reiterate their strong preference to bind themselves to the EU's climate policies, adopting the EU position that global temperatures should not climb above 2° Celsius. In addition, they continue their preference for multilateral/global action by stipulating that future international actions, post-2012, should be "made within the UNFCCC framework" (ibid., 14). The Dutch plan to focus domestically on reducing NO_x emissions includes "a system of tradable energy efficiency certificates, ... and greater energy efficiency for equipment and vehicles" (ibid., 15). Internationally, the Dutch are researching the trading of Assigned Amount Units (AAU) as an option to JI and CDM.

Another sign of the contributions the Dutch have made to the global climate change agenda was the selection of the former VROM International Affairs Director, Mr. Yvo de Boer as Executive Secretary of the UN Climate Change Secretariat in 2006. While the Dutch seem to have been rudderless for a few years, they now seem poised to resume their role as climate policy leaders. In sum, the Dutch seek to balance the norms of *Stewardship*, *Equity*, *Polluter Pays*, and the *Precautionary Principle* with *Economic Efficiency*, and are seeking a role as global leaders to promote global agreements to reduce emissions.

Conclusions

How have the Dutch constituted climate change, and thus their responses to it? How and why has their response changed? *Sustainable Development* serves as the focal point for Dutch policy, exhibiting the emergence and instantiation of an important norm, while demonstrating the discourses that have surrounded how and why the Dutch have created environmental policies and plans such as the NEPPs and Climate Memorandums and Notes. The dominant forces that have shaped Dutch identity, interests, and actions regarding climate change are summarized below.

The first conclusion is that material reality/context matters. The impact of external and internal events has buffeted the Dutch focus on climate change. The growing awareness of environmental degradation led to the first NEPP1 and the first "green election" in Europe. In addition, the introduction of the norm of *Sustainable Development* with conclusive scientific proof from an internal research institute led to a push for great change in the NEPP1. The failure of COP6 in combination with the Netherlands' failure to reduce CO_2 emissions led to a time of retrenchment and disillusionment. The veto role played by the US has served to stiffen the resolve of the Netherlands, and within its role in the European Union, to enhance emission reductions. The increasingly globalized world has prompted the Dutch to refine their approach to understanding environmental issues that are of concern to all humans. However, the economy continues to duel with the environment for prominence in the arena of Dutch politics.

At the same time, the leaders of the Netherlands have controlled the process of environmental activity. In this centralized, policy-making structure, the government has led efforts to promote *Sustainable Development*. Likewise, political parties have played a significant role, e.g., the CDA pushed for a strong policy in 1989, while the return of the CDA in 2003 has led to a reorientation of climate change policy. VROM and its ministers also have played a crucial role in defining and shaping environmental policy. For example, Minister Pronk headed the COP6 and took the failure of the meeting personally as an attack on Dutch policy. Likewise, the current State Secretary van Geel and Prime Minister Balkenende have continued to direct the environmental process during the turbulence of recent events.

The second conclusion is that the social identity of the Dutch strongly influences their actions. The Dutch perceive themselves as part of a global world and thus greatly interdependent. The government clearly perceives its approach as necessitating action at both the domestic and the international level. The Dutch are influenced by outside events and ideas but must protect themselves from the threat of climate change. While they reduce emissions domestically, they must also rely on other states to reduce their emissions as well. This is further nested within their perceived responsibility grounded in *Sustainable Development*, the *Polluter Pays Principle*, and *Precautionary Principle*, with intergenerational justice for themselves and the world. A final layer is added with their growing conceptualization that they must operate within an increasingly globalizing world, where all humans share similar problems and must share solutions.

The third conclusion is that Dutch identity is intimately connected with their interaction with nature. From a modernist perspective, they define their surroundings as something to be controlled, confident that they can overcome any challenges with technology, knowledge, and planning. In this sense, the Dutch appear to have accepted the discourse of Green Governmentality (as defined by Bäckstrand and Lövbrand in this volume). In combination, perceived results of climate change directly challenge Dutch self-conception. Their self-stated value of well-being and quality of life—even their very existence—is directly threatened by increasing pollution and sea level rise. This psychological connection with land and sea is fundamental in order to understand the magnitude of the perceived threat. Climate change, as with all other forms of environmental degradation identified in the NEPPs, directly threatens the symbiotic relationship of the Dutch with their land. It is this relationship that drives them to act to protect themselves and to improve the environmental quality of their surroundings grounded in the norm of *Stewardship*, and extending their sense of responsibility to help others in the world through the norm of *Equity*, not simply themselves.

On the surface, there appears to be a contradiction in that the Dutch have a strong social/stewardship identity with nature and the environment, and thus seek to preserve and improve their surroundings, while at the same time they highly value economic growth and are an energy-intensive, developed state. The Dutch seem to have found a compromise between *Sustainable Development* and *Economic Efficiency*. Additionally, the economic focus reflects the prominence of the discourse of Ecological Modernization as well within the Netherlands.

Clearly the Dutch have constructed a unique conceptualization of the norm of *Sustainable Development*. For example, the NEPP3 introduced the concept of "decoupling" which "refers to improving living standards (economic growth) while at the same time reducing the environmental pressure" (VROM 1998b, 16). Present even in the latest *Climate Policy Evaluation Memorandum*, the push for "absolute decoupling" signifies the search to decrease emissions alongside economic growth (VROM 2005a, 14). At times, this relationship has leaned more heavily toward the promotion of economic interests and the norm of *Economic Efficiency*, including adopting economically centered measures such as emission trading and CDM. In addition, much of the Netherlands' industrial sector has readily adopted the norm of *Sustainable Development* (with caveats) as an "opportunity" (see Paterson and Stripple chapter) in "identifying economic opportunities in climate change mitigation" (Fisher 2003, 91), as has the government with its "Clean, Clever, Competitive" proposal for the EU. While seeming contradictory, in the Dutch context it is not. It conforms quite well to their identities of merchants, stewards of nature, and pragmatists: get done what it is necessary to do to help the environment, while promoting their own economic well-being. To the Dutch, *Sustainable Development* has come to mean a fluctuating balance between economic prosperity, and sacrifices to reduce emissions and protect the environment.

In conclusion, it appears that the Dutch may be moving more toward the discourse of reform Civic Environmentalism (introduced in the Bäckstrand and Lövbrand chapter). Many voices and ideas have come to shape Dutch climate change policy. Gradually, the Dutch have had a slight conceptual shift or redefinition of climate change meaning internally and externally. They have shifted their policies to reduce domestic and international emissions in response to failures and successes, and the globalizing world. Yet, they have never abandoned their promise and responsibility to *Sustainable Development*, adopting the norm but making it uniquely Dutch. From a norm-centered perspective, two international norms have been instantiated in Dutch society and policy, and recursively, the Dutch have exported their own understandings to the global arena.

References

Andersson, M. and Mol, A.P.J. (2002), The Netherlands in the UNFCCC Process – Leadership between Ambition and Reality. *International Environmental Agreements: Politics, Law and Economics*, 2, pp. 49–68.

Andeweg, R.B. and Irwin, G.A. (2002), *Governance and Politics of the Netherlands*. New York: Palgrave Macmillan.

Biedermann, F. (1992), Cold Feet over Carbon Tax? *Chemistry and Industry*, 5 (2), p.164.

Cramer, J. (1989), The Rise and Fall of New Knowledge Interests in the Dutch Environmental Movement. *The Environmentalist*, 9 (2), pp. 101–120.

de Jongh, P.E. (1989), The Dutch Environmental Policy Plan: How to Work Together with Industry on Sustainable Development. *UNEP Industry and Environment*, 12 (3–4: Jul–Dec), pp. 20–26.

de Jongh, P.E.. and Captain, S. (1999), *Our Common Journey: a Pioneering Approach to Cooperative Environmental Management*. London: Zed Books.

Dijkink, G. and van der Westen, H. (1992), Green Politics in Europe: the Issues and the Voters. *Political Geography*, 11 (1: January), pp. 7–11.

Economist (1989a), "Straws in the Windmills." 311 (7601: May 6), p.45.

Economist (1989b), "Greening Europe." 313 (7624: Oct 14), pp. 21–22.

Election World (2004), *Elections in the Netherlands*. <http://www.electionworld.org/netherlands.htm>.

EU2004 (Netherlands European Union Presidency) (2004), *Informal environmental council*. As summarized at <http://www.eu2004.nl/default.asp?CMS_TCP =tcpAsset&id=AV2CCD9E8D1A455FB3970D9512B88791>.

European Community (1993), Towards Sustainability. *Official Journal of the European Communities* <http://ec.europa.eu/environment/env-act5/5eap.pdf>.

European Environment Agency (2003), *Greenhouse Gas Emission Trends in Europe, 1990–2000*. <http://reports.eea.europa.eu/topic_report_2002_7/en/Topic_7.pdf>.

Farla, J. (2006), "Netherlands Climate Policy." Private e-mail correspondence with Mary Pettenger, April.

Fisher, D.R. (2003), *National Governance and the Global Climate Change Regime*. Lanham: Rowman & Littlefield Publishers, Inc..

Germino, D. (2001), Meindert Fennema: Political Theory in Polder Perspective. *The Review of Politics*, 63 (4), pp. 783–804.

Hekstra, G.P. (1986), Will Climatic Changes Flood The Netherlands? Effects on Agriculture, Land Use and Well-Being. *Ambio*, 15 (6 Sept), pp. 316–326.

Hendriks, F. and Toonen, T.A.J., eds. (2001), *Polder Politics: The re-invention of consensus democracy in the Netherlands*. Aldershot: Ashgate Publishing Limited.

Hoes, O.A.C., W. Schuurmans, and Strijker, J. (2005) "Going Dutch" in *International Water Power and Dam Construction*, 57 (1: Jan.), pp. 40–44.

Johnson, H.D. (1995), *Green plans: greenprint for sustainability*. Lincoln, NE: University of Nebraska Press.

Laitner, S. and Parker, G. (2006), EU warned of looming climate catastrophe. *Financial Times*, October 20, p. 8.

Liefferink, D. (1999a), "The Dutch national plan for sustainable society" in N.J. Vig and R.J. Axelrod, eds. *The Global Environment: Institutions, Law, and Policy*. Washington, D.C.: Congressional Quarterly Press, pp. 256–278.

________. (1999b), "NL and global warming." Private e-mail correspondence to Mary Pettenger, 18 March.

________. (1997), "The Netherlands: a net exporter of environmental policy concepts" in M.S. Andersen and D. Liefferink, eds. *European Environmental Policy: the Pioneers*. Manchester, UK and New York: Manchester University Press, pp. 210–250.

Lobe, J. (2006), *Development: Netherlands Leads World in 'Quality' Aid*. International Press Service Agency, August 13. Available from: <http://www.ipsnews.net/print.asp?idnews=34329>.

Metz, B. and Vellinga, P. (1990), Views from Other Nations: The Netherlands. *EPA Journal*, 16 (2: Mar/Apr). p.38.

MEZ (2005), *Nu voor later – Energierapport 2005*. The Hague: MEA.

National Institute of Public Health and Environmental Protection (RIVM). (1989), *Concern for Tomorrow, A National Environmental Survey, 1985–2010*. Bilthoven, The Netherlands: RIVM.

NEAA (2005), *The effects of climate change on the Netherlands*. Bilthoven, Netherlands: MNP Bilthoven.

New Scientist (1992), Clean up, or else. 133 (1804: 18 Jan), p.11.

________. 1989. Dutch voters land Lubbers back in power on strength of green ticket. 123 (1682: 16 Sept), p.24.

OECD (2006), *Gross domestic product*. <http://www.oecd.org/dataoecd/48/4/33727936.pdf>.

Pettenger, M. (2000), *A Small State Constructing a Lead Role: The Netherlands and Climate Change*. Ph.D. Dissertation, University of Denver.

Reuters (2002), *Dutch, World Bank Sign First Clean Energy Deal*. <http://www.planetark.org/dailynewsstory.cfm/ newsid/ 14110/story.htm>.

Rochon, T.R. (1999), *The Netherlands: Negotiating Sovereignty in an Interdependent World*. Boulder, Colorado: Westview Press.

Schama, S. (1987), *The Embarrassment of Riches: An Interpretation of Dutch Culture in the Golden Age*. New York: Alfred A. Knopf, Inc.

Shared Spaces (2003), Interview with Pieter van Geel. 1 (April). <http://www.sharedspaces.nl/>.

Tuininga, E.J. (1994), Going Dutch in environmental policy: a case of shared responsibility. *European Environment*, 4 (4: Aug), pp. 8–13.

United Nations (1969), General Assembly Second Committee. *Agenda items 38 and 43*, 1276th Meeting, 10 November.

van der Straaten, J. and Ugelow J. (1994), "Environmental policy in the Netherlands: change and effectiveness" in M. Wintle, ed. *Rhetoric and Reality in Environmental*

Policy: The Case of the Netherlands in Comparison with Britain. Aldershot: Avebury Publishing Company, pp. 118–144.

Verhoog, W. (1999), One issue that's here to stay. *Environmental News from the Netherlands*, 2 (Apr), pp. 3–4.

Voorhoeve, J.J.C. (1979), *Peace, Profits and Principle: A Study of Dutch Foreign Policy*. The Hague: Martinus Nijhoff.

VROM (2006a), *CO2 and NOX data 2005*. <http://international.vrom.nl/pagina.html?id=10046>.

________. (2006b), *Make markets work for Climate*. <http://international.vrom.nl/pagina.html?id=9957>.

________. (2006c), *5.75% biofuels in the Netherlands by 2010*. <http://international.vrom.nl/pagina.html?id=9869>.

________. (2006d), *The Netherlands to set up climate projects with China*. <http://international.vrom.nl/pagina.html?id=10106>.

________. (2005a), *Climate policy evaluation memorandum 2005: on the way to Kyoto*. <http://www2.vrom.nl/docs/internationaal/On%20the%20way%20to%20Kyoto.pdf>.

________. (2005b), *EU demonstrates leadership in approach to climate change*. <http://international.vrom.nl/pagina.html?id=9350>.

________. (2005c), *Unique nitrogen oxide emissions trading system to come into effect on 1 June*. <http://www.sharedspaces.nl/pagina.html?id=9439>.

________. (2004a), *Sustainable Action Progress Report*. <http://www2.vrom.nl/docs/internationaal/engelse%20versie%20voortgangsrapportage%202004.pdf>.

________. (2004b), *Clean, Clever, Competitive*. <http://www.sharedspaces.nl/docs/internationaal/IEC%20040625%20Knowledge%20document%20def.pdf>.

________. (2004c), *EU environment ministers to discuss post-2012 climate policy*. <http://www2.vrom.nl/pagina.html?id=9260>.

________. (2003), *Sustainable Action*. <http://www2.vrom.nl/docs/internationaal/summary%20actionprogramma%20SD%20text.pdf>.

________. (2002), *Vaste waarden, nieuwe vormen (Set Values, New Forms) – milieubelied 2002–2006*. The Hague: VROM.

________. (2001), *Where there's a will there's a World: working on sustainable development*. (NEPP4) The Hague, The Netherlands: SDU Uitgeverij's.

________. (2000), *Netherlands' climate policy implementation plan, part II*. <http://www.minvrom.nl/minvrom/pagina.html?id=1472>.

________. (1999a), *Environmental policy of the Netherlands: an introduction*. The Hague, The Netherlands: VROM.

________. (1999b), *Netherlands' climate policy implementation plan, part I*. <http://www.minvrom.nl/minvrom/pagina.html?id=1428>.

________. (1998a), *Update of the second Netherlands' National Communication on Climate Change Policies*. The Hague, The Netherlands: VROM.

________. (1998b), *National Environmental Policy Plan 3*. The Hague, The Netherlands: VROM.

________. (1996), *The Second Netherlands' Memorandum on Climate Change*. The Hague, The Netherlands: VROM.

________. (1994a), *The Environment: Today's Touchstone*. (NEPP2) The Hague, The Netherlands: VROM.

________. (1994b), *Netherlands' National Communication on Climate Change Policies*. The Hague, The Netherlands: VROM.

________. (1989), *To choose or to lose: National Environmental Policy Plan*. (NEPP1) The Hague, The Netherlands: SDU Uitgeverij's.

WCED (1987), *Our common future*. [Brundtland Report] Oxford and New York: Oxford University Press.

Winsemius, P. (1990), *Guests in our own home: thoughts on environmental management*. McKinsey & Company. Translation of *Gast in Eigen Huis*. Alphen aan den Rijn: Samsom H.D. Tjeenk Willink.

Zito, A.R., Brückner, L., Jordan, A. and Wurzel, R.K.W. (2003), Instrument innovation in an environmental lead state: 'new environmental policy instruments in the Netherlands'. *Environmental Politics*, 12 (1), pp. 157–178.

Chapter 4

The Rise of Japanese Climate Change Policy: Balancing the Norms of Economic Growth, Energy Efficiency, International Contribution, and Environmental Protection

Takashi Hattori

Introduction

This chapter examines the development of Japanese domestic norms on climate change policy from the late 1980s to the early 2000s. How and why did Japan define, adopt and act on these norms, and which domestic actors have shaped the norms? Since the emerging salience of the climate change issue in the international scientific community in the 1980s, Japan has shaped and reshaped its domestic climate change policy. This process is closely linked to the struggle between the often competing norms of economic growth, energy efficiency, international contribution, and environmental protection in Japan. The chapter begins with a short description of each norm, and then traces the three stages of norm diffusion in Japanese policy.

Japanese Norms

The phase of economic recovery and growth in Japan after World War II is often called the *Sengo* (after the war) period. Sixty years after the fact, *Sengo* is still continuously used to describe the time since WWII. The *Sengo* era began with the slow economic recovery of the 1950s, which was followed by rapid industrial development in the 1960s. The economic growth was also shaped by the Japan–United States Security Treaty, as Japan, being allowed under the terms of the treaty, did not rebuild its military power in the era of the Cold War but concentrated on economic development. During this time period the norm of economic growth began to dominant in Japan, ensuring that the country's limited resources were allocated towards economic prosperity. One significant norm entrepreneur during this phase was the Ministry of International Trade and Industry (MITI); however, its role has often been debated (Johnson 1982; Friedman 1988; Huber 1994; Callon 1995). The role of MITI will be further described below.

The 1970s saw the rise of a new norm, energy efficiency, in Japan emerging out of the two significant energy crises Japan experienced. Because Japan relies on oil imports to meet much of its energy needs, it was severely impacted by the imposed oil shortages and price increases of 1973 and 1979 which caused recession and social turmoil. In response, MITI created a new branch, the Natural Resources and Energy Agency (NREA), in 1973. In addition, there was another significant change demonstrated by "The Law Concerning the Rational Use of Energy" which was introduced in 1979. Following these adjustments to the energy crises, Japanese industries were required to rationalize their energy consumption. Thus, the domestic norm of pursuing energy efficiency began to develop in this period. In addition, by the late 1980s (when climate change became part of the political agenda), Japan had become the most energy-efficient, developed country.

The third norm, international contribution, also emerged in the context of the *Sengo* period. After reaching an industrially developed stage, Japan sought to make a greater contribution to world politics (Mervio 2005; Ohta 2005; Kameyama 2003; Kawashima 1997a, 1997b). Having a constitution that renounced war and included the abandonment of military forces, Japan sought an affirming national agenda within the international arena in the late 1980s. At this time, a new domestic norm, international contribution, thus emerged, which illustrated a normative desire to validate and promote a leadership role by Japan in international politics. In the late 1980s, international contribution was a norm with which Japan could assert to other nations that Japan could contribute to the international community in a peaceful manner and overcome the history of WWII. As the Cold War drew to a close, the international focus expanded as well to include global environmental issues. Consequently, Japan began to emphasize global environmental issues as part of its foreign policy and its worldview. This fit well to its national identity of a peaceful nation.

In contrast, actions to fulfill this norm in a military manner have been controversial in Japan. In the 1990s, Japan could not send personnel from its Self-Defense Forces abroad during the Gulf War. However, in the 2000s, Japan chose to send its Self-Defense Forces to support the international efforts for the post-war Iraqi reconstruction. There were powerful domestic debates on how to make an international contribution in this issue and abide by the Japanese constitution. Compared to such an issue, the global environment has been an issue within which Japan can consider international contribution much more easily.

Finally, the norm of environmental protection in Japan has played an integral role in shaping and reshaping Japan's actions. The islands of Japan are surrounded by seas and endowed with forests in the mountains and rice fields in the plains. Some would argue that the Japanese have been nature lovers since ancient times. Traditionally, the Japanese have been fond of nature, where they have even found spirituality as it is often said that Japan has "eight million gods." After the Meiji Revolution in 1868, however, *Fukoku Kyohei* (Rich Nation, Strong Army) became a slogan for the nation's development, and nature came to be considered less important. After WWII, *Kyohei* (the Strong Army) did not return, but the concept of *Fukoku* (Rich Nation) dominated other norms in Japan for a time. Thus, the norms of economic growth and environmental protection have been in continual conflict within Japan for over one hundred years.

However, the norm of environmental protection has gradually risen in stature. For example, in the late 1960s, Japan faced a serious side-effect of rapid industrial growth called *Kogai* (public nuisance). People suffered severe health problems caused by unregulated air and water pollution from factories in industrial sectors. The Diet (the Japanese parliament—*Kokkai*) responded in 1970 by passing laws to alleviate *Kogai* (this Diet session was specially named *Kogai Kokkai*), and most significantly the establishment of the Environment Agency in 1971. In addition, during the 1980s, Japanese companies operating overseas faced criticism that they lacked environmental consciousness, and Japanese official development assistance was causing environmental degradation such as the logging of tropical forests. In response, Japanese companies and the government reacted by becoming more environmentally conscious to overcome such criticism (Barrett 2005), rather than ignoring the rebuke. This shift demonstrates the importance the Japanese held for environmental protection in contrast to economic growth.

By the 1990s, Japan had shifted to even more environmentally conscious policies and laws, including the Basic Environment Law in 1993 and the Environmental Impact Assessment Law in 1997. Some have argued that the decade of the 1990s was Japan's period of ecological modernization (Barrett 2005). However, the environment has continued to gain greater significance as seen by the theme of the 2005 World Exhibition (held in Aichi, Japan) of "Nature's Wisdom." The domestic norm of environmental protection has re-emerged.

Japan, as an island country, which has affluence in mountains, rivers, and seashores, but has little natural resources for its industrial development, has struggled to re-build itself as a peaceful nation which puts emphasis on the environment. The norms of economic growth, energy efficiency, international contribution, and environmental protection have importance in Japan and they are the most significant norms shaping climate change policy making. The chapter now turns to discuss Japanese climate change policy.

Domestic Norms and Climate Policy

The development of Japanese domestic norms on climate change policy can be divided into three stages. The first stage is the period in which the United Nations Framework Convention on Climate Change (UNFCCC) was created from the late 1980s to 1992. The second stage is the period in which the Kyoto Protocol was formulated and implemented from 1992 to the late 1990s. The third and current stage is the post-Kyoto negotiation period beginning in the early 2000s.

Many social construction scholars emphasize the creation of international norms; yet, as noted in the introduction to this book, examination of the formulation of domestic norms is rare (Katzenstein, Keohane, and Krasner 2002; Finnemore and Sikkink 2001; Wendt 1999; Checkel 1997). This chapter focuses on the shaping and reshaping of domestic norms, and draws theoretical inspiration from Hopf (2002), who illustrated a social cognitive structure. In his framework, one norm overwhelms others in changing conditions over time. Following his example, this chapter examines various domestic norms and illustrates how certain domestic norms were shaped and reshaped over time. As a case study using Japan, I examine the emergence and transformation of norms on climate change policy at each stage. Throughout this analysis, I identify the socially constructed realities of climate change and how these realities have been constructed in a uniquely Japanese context.

Stage 1: Period of Formulation of the UNFCCC

The first stage takes place during the period of formulation of the UNFCCC from the late 1980s to the early 1990s. Japanese focus during this time was on identifying global warming as a global environmental issue. Primary domestic agents included the ruling Liberal Democratic Party (LDP) and the relevant ministries (the Environment Agency and MITI). The rise in salience of global warming did not proceed smoothly within Japan. In the beginning, Japan did not commit to the "CO_2 Emission Reduction Commitment Norm" (see the chapter by Cass in this volume) at the Noordwijk Conference in 1989. However, even with this sign of limited commitment, the Japanese media reported that the conference produced a political statement without numerical target commitments. This incident increased public awareness of the international pressure to produce emission reduction targets. However, the Japanese government failed to make an international contribution at this time, because Japan had not yet embraced the norm of international contribution in relation to climate change.

However, the norm of environmental protection had already begun to rise in salience. And in the late 1980s, the development of the domestic norm of international contribution in Japan began to be strongly linked to global environmental issues. Public concerns about global environmental issues were heightened. For example, the Group of Seven (G7) Leaders' Summit, held in Lyon, France in 1988, was called the *Kankyo Samitto* (the Environment Summit) by the Japanese media. *Kankyo Zoku* (the Environment Tribe) was the name given to a political group promoting the awareness of global environmental issues in the ruling LDP.

The diffusion of the norm of environmental protection was also promoted by a principal norm entrepreneur, who linked the global environment issues with Japanese international contribution. Prime Minister Noboru Takeshita became pro-environment and encouraged increasing Japanese international contributions to global environmental issues. For instance, in partnership with the United Nations Environment Program (UNEP), Japan hosted the Tokyo Conference on Global Environmental Protection to discuss possible international cooperation on environmental issues, because the LDP "had given a big boost to the Environment Agency" (Kawashima 2000, 42).

Another environmentally promoted event occurred in the late 1980s. The Cabinet, the supreme decision-making forum for the administration led by Prime Minister Takeshita, established the Council of Ministers for Global Environmental Conservation on 12 May 1989. Various Councils of Ministers in Japan are established by oral understanding at meetings of the Cabinet to discuss issues at the ministerial level. The Council represents a preliminary stage within the Cabinet's decision-making process and provides a forum for ministerial level discussion of policies and measures adopted (or under consideration) by the Cabinet. The Council of Ministers for Global Environmental Conservation was established in the aftermath of the G7 *Kankyo Samitto* in 1988. The Council was an ad hoc entity aimed at achieving close cooperation among ministries and agencies to devise effective and comprehensive measures for addressing serious threats to the global environment.[1] Notably, Ryutaro Hashimoto, who became the Prime Minister during the Third Session of the Conference of the Parties (COP3) to the Kyoto Protocol, was then the LDP's first secretary.

In addition, under political pressure from the ruling party, the relevant ministries in the administration prepared "The Action Plan to Prevent Global Warming." The plan was released on 23 October 1990 and contained policies and measures to combat global warming, including numerical targets for reducing CO_2 emissions. The plan covered various measures to limit global warming, with policy objectives to stabilize per capita CO_2 emissions and total CO_2 emissions (if technological development were to progress more than expected) at the 1990 level by 2000. The plan was the result of the formulation of the Cabinet-level organization, and was adopted at this Council.

The growing influence of the norm of international contribution was clear. For example, Kawashima (1997a, 1997b) interviewed ten experts in academia, government and research institutes who had been deeply involved in the issue in Japan from April to October 1993. She found that the majority of the interviewees concluded that international politics (variously perceived as "international contribution" or "international cooperation") influenced Japanese policy-making.[2] Thus, there was an apparent confluence of two domestic norms, the norm of environmental protection was expanded to include global environmental issues at this stage, and to include the norm of international contribution. The *Environmental Law Journal* (October 1991, No. 19) featured a volume entitled "Global Environmental Problems and

---

1 The revised version of the oral understanding at the meeting of the Cabinet regarding the Council of Ministers for Global Environmental Conservation can be found at <http://www.kantei.go.jp/jp/singi/kankyho/kankyo.html>.

2 Kameyama (2003) revisited the reasoning of international contribution to climate change policies during this period. She also illustrated further development after the period. See also Mervio (2005) and Ohta (2005).

International Response." The Journal was edited by the Japan Center for Human Environmental Problems, which wrote:

> It is often said that "*Heisei Gannen Wa Chikyukankyo Gannen*". This means that the first year of the Heisei—an Imperial era that dates from 1989 to the present—was the first year of the global environment, a major turning point for environmental policy in Japan, not only among central and local governments but also among people and companies. At this time a common consciousness developed that policies to protect the earth were a priority issue for people, and thus the global environmental era began (Japan Center for Human Environmental Problems 1991, 129).[3]

The diffusion of these norms also extended to the general public. In March 1990, the Prime Minister's Office conducted a public opinion poll on *Chikyu Kankyo Mondai* (the global environmental problem). The percentage of people who said the "global environmental problem" was an environmental problem of concern more than doubled from 20.7 percent in the October 1988 poll to 42.4 percent in the March 1990 poll. In addition, 59.7 percent indicated that global environmental problems were "the first priority among the world's problems." Finally, 88.9 percent answered that they were aware of the issue and did something in their daily lives to contribute to improving the global environment (Japan Center for Human Environmental Problems 1991, 153–158). Additionally, the salience of global environmental problems, including global warming, spread throughout the Japanese public, industry, and non-governmental organizations (NGOs) during this stage (Sato 2003).

In short, the norms of economic growth and energy efficiency had begun to be balanced with the norm of environmental protection and international cooperation by the 1990s. For example, representing Japanese industries, Keidanren (the Japan Federation of Economic Organizations) enacted the Global Environment Charter in April 1991. Its declared guidelines for corporate actions, in response to global problems, stated that "Companies shall cooperate in scientific research on the causes and effects of such problems as global warming ... Companies shall actively work to implement effective and rational measures to conserve energy and other resources, even when such environmental problems have not been fully elucidated by science" (Keidanren 1991).

Nevertheless, the tension between the norms of economic growth and energy efficiency, and the norm of environmental protection was highlighted during the inter-ministerial drafting process of the Action Plan for the Council of the Ministers. Within the government, the Environment Agency (EA) insisted on stabilizing CO_2 emissions, whereas MITI asserted that such a goal was impossible to achieve; this disagreement resulted in a compromise of target levels in the Action Plan. The EA, using a simple model that summed potential methods of energy conservation, insisted that stabilization of CO_2 was achievable. MITI argued, on the basis of the Long-Term Energy Plan, that CO_2 emissions would increase by 16 percent by the year 2000 compared with the 1988 level. Therefore, at a minimum, an 8 percent

3 Translated for this chapter by Takashi Hattori.

increase in emissions would be unavoidable. After taking a reduction of 2 percent in the transportation sector (estimated by the Ministry of Transportation) into account, MITI proposed a 6 percent increase, insisting that, without technical innovation, it would not be possible to stabilize CO_2 emissions at the 1988 level by 2000. The EA and MITI compromised (taking the expected population growth of 6 percent into account) by making two targets: one based on per capita CO_2 emissions and the other on CO_2 emissions under the condition of realizing unexpected technical innovation (Schreurs 2002; Kawashima 2000). In addition, MITI and its NREA insisted on their own estimate of energy supply and consumption that would allow expected economic growth. The EA asserted that there were ways to save energy, but it failed to convince MITI. The two agreed only on a compromise of wording by using "stabilization" of CO_2 emissions.

The activities of environmental NGOs were limited at this stage. Three Japanese environmental NGOs, Friends of the Earth Japan, Greenpeace Japan, and Citizens' Alliance for Saving the Atmosphere and the Earth (CASA), participated in the Second World Climate Conference in October 1990. At the conference, Friends of the Earth selected participants from the Japan branch to join with a group of members from developing countries, such as Ghana and Kenya, to discuss the following questions: "How influential are NGOs to the government?" and "How highly does the government evaluate NGOs" (Takeuchi 1998, 57)? In this discussion, the Japanese NGOs had little impact.

Japan's policy on global warming however had become more pronounced. By producing the national commitment for a numerical target, Japan was ready to contribute further to international regime formation; thus, Japan stepped into the international arena with an emission reduction proposal in the very earliest negotiations formulating the UNFCCC. The Intergovernmental Negotiation Committee (INC), under the guidance of the United Nations, assumed responsibility for formulating a convention on climate change. The international negotiation began in February of 1991 with the first meeting of the INC and continued to May 1992, the second half of the fifth meeting of the INC held just before the Rio Earth Summit (Mintzer and Leonard 1994; Paterson 1996). Japan presented the "pledge and review" proposal at the Second Session of the INC, asking each country to pledge its own target goal for emissions and to allow other members to review its progress toward the goal (Akao 1993).

In summary, the first stage of Japan's development of climate change policy revealed the often conflicting norms of international contribution, economic growth, energy efficiency, and environmental protection which all contributed to a legitimized national policy that considered the climate change problem as one of a number of important global environmental issues. With a strong economic background and a norm for international contribution, Japan was able to succeed in raising civic awareness and participation while enlisting business, and in so doing, formulated a national target for carbon dioxide emissions upon which the Japanese international proposal was based. In addition, the Japanese domestic norms flowed outward to the international level in the process of encouraging other states to adopt the "pledge and review" proposal.

Stage 2: Period of Formulation and Implementation of the Kyoto Protocol

The second stage of norm development was the period during the late 1990s in which the Kyoto Protocol was formulated and implemented. In this time period climate change policy in Japan was the sole important environmental issue. Primary agents in this stage included the political parties, the Cabinet Secretariat, the EA, and MITI, as well as the private sector, the Keidanren, and a growing number of environmental NGOs.

The first norm that significantly influenced Japanese foreign and domestic policy during this stage was that of international contribution. At the UNFCCC's First Session of the Conference of the Parties (COP1) in 1995, Japan presented its readiness to host the Conference of the Parties (COP) at the third session or later. At COP2 the following year, Japan officially agreed to host COP3. In addition, Prime Minister Ryutaro Hashimoto attended the Denver Summit and the Special Session of the United Nations General Assembly in 1997. During these meetings he found a large gap between the European Union (EU) and the United States in terms of their stances on how to set the level of emission reduction of greenhouse gases. After returning to Japan, he instructed the Cabinet Secretariat to coordinate the relevant ministries to make further efforts, both domestically and internationally, to ensure the success of the forthcoming COP3 (Hattori 1999).

Prime Minister Hashimoto, in July 1997, also instructed the Cabinet Secretariat to coordinate and propose a comprehensive strategy for the Kyoto Protocol negotiations. Specifically, the Prime Minister instructed the EA, MITI, and the Ministry of Foreign Affairs (MoFA) to coordinate for this purpose and also asked them to consider creating a joint conference of relevant advisory councils.[4]

The Ad-hoc Group on the Berlin Mandate (AGBM) negotiations conducted eight meetings between August 1995 and October–November 1997, and its negotiating documents were to be finalized at COP3 in December 1997 (Grubb 1999). To move the international negotiations further along, Japan submitted a proposal to the AGBM in October 1997 that asked developed member countries to reduce various emissions by 5 percent from the base level (Tanabe 1999). The proposal[5] called for a base reduction rate for all developed countries that would reduce carbon dioxide (CO_2), methane (CH_4), and nitrous oxide (N_2O) emissions to 5 percent below 1990 levels by 2008–2012. It also proposed a differentiation formula that would vary the level of reduction on the basis of each country's economic strength.

This proposal represented a compromise of the inter-ministerial consultations among ministries and agencies representing different domestic norms in Japan. The EA had initially proposed a 7 percent reduction on the basis of its analysis by the

4 Prime Minister's instruction, 15 July 1997. See Hattori (1999).

5 Press Conference by Chief Cabinet Secretary Kanezo Muraoka on 6 October 1997.

Asia-Pacific Integrated Model (AIM). It claimed that a substantial reduction was necessary to maintain the global climate and, further, that the reduction would be feasible. Conversely, MITI consulted with other ministries such as the Ministry of Transportation (MoT) and Ministry of Construction and then proposed stabilization at the 1990 level. The Ministry of Foreign Affairs (MoFA) insisted on a 5 percent reduction in order to frame an internationally acceptable rate of reduction between the European Union's 15 percent reduction proposal and the unknown level of reduction offered by the United States.

Under the coordinating efforts by the Cabinet Secretariat, the EA, MITI, and MoFA agreed to take the following three steps. First, the Japanese position shifted on which GHGs to include. Japan shifted from one gas (CO_2) to three gases, thus adding a 0.5 percent reduction. Second, they introduced differentiation which favored Japan because of its already-achieved, high level of energy efficiency, allowing a further 2.5 percent reduction. Third, Japan sought flexibility on non-compliance, which could act as a buffer for the 2 percent reduction (Hattori 1999).

By this time, the influence of the norms of environmental protection and international contribution had nudged Japan to single out climate change policy from the other various global environmental issues. This shift also percolated from the government to the general population. The Prime Minister's Office conducted a public opinion poll on *Chikyu Ondanka Mondai* (the global warming problem) in June 1997.[6] Of the respondents, 79.4 percent were "interested in the global warming problem" (25.3 percent answered "very much," and 54.1 percent "to some extent"), whereas 82.2 percent of respondents indicated they were "concerned about global warming" (27.7 percent answered "very much" and 54.5 percent "to some extent"). In response to the question, "Which do you think was more responsible for causing global warming, the lifestyle of individuals or industrial activities?" 10.2 percent of respondents answered "the main cause is individuals' lifestyles," 29.5 percent answered "mainly caused by lifestyle, but also by industrial activities," 43.9 percent answered "mainly caused by industrial activities, but also by lifestyle," and 10.6 percent answered "the main cause is industrial activity." Nearly half (49.7 percent) of the respondents indicated that actions by governments and international organizations against the global warming problem were "the first priority among the world's problems."

At the governmental level, additionally steps were taken to prepare for COP3. For example, the establishment of the Joint Conference of Relevant Advisory Councils on Domestic Measures for Addressing Global Warming was based on the Prime Minister's Decision. The Decision stated that, "[because] the measures cover a broad range of policy areas, to effectively implement the various measures it is necessary for the whole government to promote coordinated and systematic linkages among the various measures" (Article 1 "The Establishment of the Joint Conference" 1997). The members of the Joint Conference were three environmental experts, five other experts, six industrial leaders, two representatives of labor and consumer groups,

6 "*Chikyu Ondanka Mondai ni kansuru Yoron Chosa*." June 1997.

and two people active in the field of mass communications. Members were selected from the membership of the relevant advisory councils.[7] The Joint Conference produced a report in November 1997 entitled "Basic Direction for Measures for the Prevention of Global Warming, Focusing on Comprehensive Measures to Restrain Energy Demand." The report concluded that, if proper measures were taken, "carbon dioxide emissions from energy consumption will drop by 2010 to virtually the same level as in 1990" (Joint Conference 1997).

Of significance to the focus of this paper, the different domestic norms in Japan were tested in the discussions of the Joint Conference. The norms of energy efficiency and economic growth were examined and slowly incorporated with the norm of environmental protection. In addition, by the time of the December 1997 COP3 meeting, Japan's norm of international contribution was strongly in place. The support for this norm was bolstered as well when Japan achieved the goal of acceptance of its proposed differentiation of reduction rates among the EU, the United States, and Japan. Since the projected absorption by sinks met the targeted reduction rate, Japan accepted a 6 percent reduction from the 1990 level in the first commitment period of 2008 to 2012. The projected reductions were –2.5 percent for carbon dioxide, methane, and nitrogen dioxide; +2.0 percent for hydrofluorocarbons (HFC), perfluorocarbons (PFC), and sulfur hexafluoride (SF_6); and –3.7 percent for carbon sinks. Japan contributed to the international negotiations by adopting its then perceived maximum reduction capacity to secure an agreement at the Conference.

Efforts in support of the dominant norms continued even after COP3, the Cabinet, led by Prime Minister Ryutaro Hashimoto, established the Global Warming Prevention Headquarters in December 1997. These Headquarters, established by a written decision of the Cabinet, are consultative bodies established to facilitate the implementation of comprehensive and effective policies by allowing multiple governmental entities to work together. The Global Warming Prevention Headquarters were tasked with promoting comprehensive, concrete, and effective measures to prevent climate change and implementing the Kyoto Protocol.[8] Through ministries and agencies, the Headquarters gathered policies and measures related to preventing climate change and adopted the "Guidelines of Measures to Prevent Global Warming" in June 1998. The Guidelines targeted a 6 percent reduction of GHG emissions from the 1990 level by 2010.

The domestic will supporting the success of Japan in hosting COP3 was strong in many contexts. Several examples serve to display the fervor of this domestic will and the importance placed on climate change as the most important environmental issue. First, representatives of political parties formed the "COP3 Team" to pressure

7 These include the Central Environment Council, Advisory Committee for Energy, Industrial Structure Council, Economic Council, Social Policy Council, Building Council, Road Council, Transport Policy Council, and Telecommunications Council. See Hattori (2000).

8 The revised version of the Cabinet's Decision on the Global Warming Prevention Headquarters can be found at <http://kantei.go.jp/jp/singi/ondanka/konkyo.html>.

the administration to make contributions at COP3 internationally and to promote environmental awareness within the Japanese society. There is no Green Party in Japan; however, during this time all the parties (both ruling and in opposition) favored promoting the prevention of climate change.

Second, in October 1998, the Diet passed the Law Concerning the Promotion of the Measures to Cope with Global Warming. The law outlined the responsibilities of the central government, local governments, companies, and the general public. It required that the central government write a basic guideline under the law, with measures that addressed emissions of each of the six greenhouse gases. In addition, the November 1997 report of the Joint Conference pointed out that "*Chikyu Ondanka* (global warming) is a serious issue directly linked to the lives of present and future generations of mankind."

Third, business sectors also desired to be seen as complying with the norm of environmental protection while, at the same time, preferring voluntary actions to government intervention in business activities In July 1996, the Keidanren announced the "Keidanren Appeal on the Environment: Declaration on Voluntary Action of Japanese Industry Directed at Conservation of the Global Environment in the 21st Century." In the appeal, the Keidanren called for the need to "prepare industry-wide voluntary action plans incorporating definite goals and steps toward enhancement of energy efficiency, and periodically reviewing the progress of such actions" (Keidanren 1996).

Fourth, the number of environmental NGOs in Japan had increased at this stage, and they gradually increased their capacity to take action. The Kiko Forum was established in December 1996 as an umbrella organization of Japanese environmental NGOs to coordinate joint activities before and during COP3. NGOs that participated in COP1, such as Greenpeace Japan, World Wildlife (WWF) Japan, People's Forum 2001, the Japan Scientists' Association, and the Citizens Alliance for Saving the Atmosphere and the Earth (CASA) became the main promoters of the Kiko Forum. The organization issued periodic newsletters and announced appeals (Kiko Forum/ Kiko Network 1998).[9] In November 1997, CASA, one of the core members of the Kiko Forum, distributed its own proposal to reduce CO_2 emissions. The CASA report concluded that it would be possible to reduce CO_2 emissions by 21 percent in 2010 compared with the 1990 level if production and consumption were maintained or reduced, and photovoltaic and wind power generation were properly introduced (CASA 1997).

In addition, during this period, the norms of energy efficiency and economic growth were closely linked to the prevention of climate change. To prepare the discussions for the Joint Conference before COP3, MITI and its NREA promoted the introduction of a "top-runner" approach to improve the energy efficiency of vehicles and consumer equipment to levels above those of the highest-level products being sold on the market. The Diet passed the revised Law Concerning the Rational Use of Energy in June 1998, which established tougher energy efficiency standards

9 The Kiko Forum was succeeded by the Kiko Network in April 1998.

for equipment. Manufacturers were obliged to surpass a weighted average value for all their products per category for each predetermined target year. The proposal also supports the norm of economic growth as well as these measures would serve Japanese companies in preparing them to set the pace in energy-efficiency in international trade.

The convergence of domestic norms in Japan all based on the need to take actions against the climate change problem was further fueled by labeling the issue "Kyoto." An ancient and revered capital of Japan, Kyoto was not only the place where the international conference at which the climate change protocol was adopted, it also became a national identity for the public's growing attachment to tackling the climate change problem during this period. In summary, the dominance by the norms of economic growth and energy efficiency was diminished during the discussions of the Joint Conference and the process of formulating the Joint Conference Report before COP3. In contrast, the domestic norm of environmental protection, tied to the norm of international contribution, gained power as the agreement of the Kyoto Protocol was achieved and the guidelines to implement the Kyoto target were set.

Stage 3: Period of Discussion of the Post-2012 Regime

The third stage is the period of discussion of the post-2012 climate change regime from the early 2000s to the present. During this time, there has been significant reframing of the climate change issue, and further contestation of norms. Domestic support in Japan for tackling the global warming problem has been challenged by the use of a seemingly more modest term, *Kiko Hendo* (climate change), and there is an emerging balance between the environment and the economy. Internationally, Japan continued to promote the climate regime and to urge the United States to ratify the Kyoto Protocol while, at the same time, working to determine the agreements for implementing the Kyoto Protocol in COP6bis and COP7 (Hamanaka 2006). Primary agents at this stage include the Ministry of Environment (MoE),[10] the Ministry of Economy Trade and Industry (METI), the Nihon-Keidanren, and environmental NGOs.

This time period also witnessed the rise of a new government in Japan with the Cabinet in April 2001, led by Prime Minister Junichiro Koizumi, starting its administration. However, support for the Kyoto Protocol did not end with the new administration. In April 2001, the Diet, both the House of Councilors and the House of Representatives, unanimously voted on resolutions to urge the government to ratify the Kyoto Protocol. Additionally, input from NGOs to parliamentary members helped in the drafting of the resolutions (Schroder 2003).

---

10 On the basis of the organizational reform conducted by the Hashimoto Cabinet, the Environment Agency became the Ministry of the Environment, and the Ministry of International Trade and Industry became the Ministry of Economy, Trade and Industry in January 2001.

However, increasing strains in the normative framework soon became apparent. In June 2001, before the Japanese Government had ratified the Kyoto Protocol, the Chairman of the Keidanren, Takashi Imai, expressed his organization's "Request for Calm and Patient Negotiations on the Issues of Global Warming." He urged the government to continue its efforts to establish an international framework that included Japan, the United States, and Europe. He also urged industries not to slow down their action against global warming, and he urged the public to have "calm discussions." He stated that, "There is a movement among the public to urge the government to ratify the Kyoto Protocol without the participation of the United States. However, this would not be enough for considering truly effective measures against global warming" (Keidanren 2001).

The tension between the norms of economic growth and environmental protection grew as well when the inter-ministerial consideration started to revise the "Guidelines of Measures to Prevent Global Warming," which had been finalized in March 2002. After the consultation between MoE and METI, the Global Warming Prevention Headquarters inserted new features into the guidelines.

First, it clarified the approach of addressing global warming by pursuing environmental and economic issues together. Second, it detailed a step-by-step approach, by dividing the lifespan of the Protocol into three periods (2002–2004, 2005–2007, and 2008–2012). Third, it stated efforts toward global participation. Fourth, it employed the Kyoto mechanisms as effective tools to pursue its objective. The revised guidelines reaffirmed the responsibilities of ministries and agencies (which had been determined during the process of establishing the original guidelines) and postponed the introduction of unexamined measures such as environmental taxes and domestic emission trading schemes.

Meanwhile, Japan ratified the Kyoto Protocol in June 2002. To implement the Kyoto Protocol, the Diet passed the "Bill on the Amendments to the Law Concerning the Promotion of the Measures to Cope with Global Warming." The revised law required that the government adopt the Kyoto Target Achievement Plan to achieve the 6 percent emissions reduction commitment called for under the Kyoto Protocol. At the same time, the revised law also *legally* established the Global Warming Prevention Headquarters[11] for the purpose of drafting the Kyoto Target Achievement Plan and supervising its implementation. In June 2002, to strengthen energy efficiency requirements, the Diet also passed the "Bill to Revise the Law Concerning the Rational Use of Energy."

During this stage, following the ratification of the Kyoto Protocol, the contestation between the four norms increased even more. These struggles frame the discussions of the future framework of climate change policy and the appropriate emission reduction commitment. For example, ministry interests can be observed through the relevant advisory councils. Advisory councils began producing reports to their

---

11 The Global Warming Prevention Headquarters had been operated on the basis of the Cabinet Decision, but it gained the legal basis for operation by this revision of the Law approved by the Diet.

ministers (and to the public) regarding the future framework on climate change after the first commitment period of the Kyoto Protocol. The Councils began to churn out numerous reports. The first report was prepared by the Industrial Structure Council in July 2003. The second was prepared by the Central Environment Council in January 2004.

In addition, international events dampened domestic enthusiasm. At COP8 in October 2003, Japan and the EU failed to launch the formal negotiation of the future framework. The Delhi Ministerial Declaration on Climate Change and Sustainable Development (Decision 1/CP.8) merely states that, "Parties should promote informal exchange of information on actions related to mitigation and adaptation to assist Parties to continue to develop effective and appropriate responses to climate change." This informal exchange of information was hopefully to have the effect of promoting domestic consideration of concrete ideas on the future framework, owing to the limited efforts underway at the international level.[12]

Domestically, the report process continued. The Global Environment Sub-Committee of the Industrial Structure Council (GESISC) produced its provocative interim report, entitled "Perspectives and Actions to Construct a Future Sustainable Framework on Climate Change," in July 2003. The sub-committee, headed by Yoichi Kaya, a professor emeritus at the University of Tokyo, consisted of 29 members from academia, industry, consumer and other groups, and labor unions.[13] The report stated that the Kyoto Protocol currently deals with only approximately one-third of the world's GHG. It emphasized that the commitment was set so as to reduce emissions, but suggesting that there are differences between states in the relative difficulty of achieving reduction targets and, further, that targets favor the EU.

In addition, the report presented "Perspectives and Actions to Construct a Future Sustainable Framework," including the following factors and concepts related to climate change: (1) the need for a technical breakthrough, (2) the diversified agenda in each nation, region, and sector, (3) the tremendous global cost of remediation, and (4) remaining scientific uncertainty. Basic concepts for a sustainable framework were identified as (1) a focus on technological solutions, (2) simultaneous achievement of effectiveness, efficiencies, and equity, (3) contribution to both economy and environment, and (4) multi-stakeholder participation.

Finally, the report proposed a "multi-faceted approach" and a "major emitters' initiative." The multi-faceted approach was defined as an approach in which "various participants will discuss various measures and accumulate one by one the actions for each field and sector." By using this approach, the report asserts:

> governments will be able not only to negotiate for treaties and protocols, but also to conduct international coordination in broader areas of regional, bilateral, and other levels. Similarly, and in parallel to such governmental actions, industries, NGOs, and individuals,

---

12 Before COP8, notable efforts were conducted by the International Energy Agency (IEA 2002), see also Baumert (2002).

13 From October 2002 to July 2003 the Sub-Committee examined the issue of the future framework on climate change.

respectively, can build their own feasible international agreements and commitments (GESISC 2003, 57).

The major emitters' initiative was defined as "[an] approach in which major emitting countries take an initiative, with authority and responsibility, to discuss measures for reduction of greenhouse gas emissions. ... [and] Major emitters have the responsibility to lead the discussion on the future international framework, and need to present a sustainable system" (ibid., 57–58). Therefore, the interim report recommended considering the involvement of the countries that emit the most greenhouse gases. Furthermore and significantly, it suggested that, "[a] creative and more practical approach in terms of the consensus-building process is required in future international discussion." (ibid., 58). In sum, the report was strongly in support of effective international measures to address climate change and thus supported the norms of environmental protection and international contribution.

On the other hand, the Global Environment Committee of the Central Environment Council produced its interim report entitled, "Climate Regime Beyond 2012 Basic Considerations" in January 2004. The Committee, chaired by Naohito Asano, a professor at Fukuoka University, consisted of 40 members from various sectors. The report pointed out the following basic goals in approaching the issue of the future framework:

(1) Maintain progress towards meeting the ultimate objective of the UNFCCC,
(2) Bring the Kyoto Protocol into effect and fulfill commitments,
(3) Achieve global participation,
(4) Ensure equity based on the principle of common but differentiated responsibilities,
(5) Build on existing international agreements through negotiation,
(6) Build international consensus by national governments with the participation of various participants, and
(7) Make the environment and economy mutually reinforcing.

Significantly, this report balances the norms of environmental protection and international contribution, with economic growth (see goal 7 above).

Both reports advocate the development of the future framework on climate change. The main difference between the two reports is how to consider the existing framework of the UNFCCC and the Kyoto Protocol. The former asserts an alternative approach to the major emitters' initiative and the latter promotes building onto existing international agreements as a basis for negotiating the climate regime beyond 2012.[14]

The difference of meaning from global warming (*Chikyu Ondanka*) to climate change (*Kiko Hendo*) in the Japanese language is worth paying attention to. *Kiko Hendo* sounds more modest, neutral, and scientific. At the third stage, where

14 Additional information on future climate change frameworks can be found in Takamura and Kameyama (2005).

stakeholders have become more aware of the consequences, both reports utilize the word, *Kiko Hendo*, in conceptualizing a future international framework.

Furthermore, there has also been private-sector discourse on "post-Kyoto" and "post-2012." For example, in November 2003, the new Nihon-Keidanren (the Keidanren and Nikkeiren [Federation of Employers Association] were merged in May 2002) used the term "post-Kyoto Protocol" in its appeal. On the other hand, WWF Japan issued its proposal using the term "post-2012" at COP10 in December 2004. Each group was focused on the emerging negotiations, but with different interests, economics versus the environment.

International and domestic events also increasingly influenced each other during the period. While the Kyoto Protocol came into force in February 2005, early on the Japanese began to implement measures in compliance with their ratification. The 2002 revised "Guidelines on Measures to Prevent Global Warming" set out a comprehensive review of policies and measures in 2004, and the 1998 Law Concerning the "Promotion of the Measures to Cope with Global Warming" was designed to create the "Kyoto Protocol Target Achievement Plan" (Article 8). The Plan, which incorporated the results of the review, was approved by the Cabinet in April 2005 (Government of Japan 2005).[15] In following the Guidelines (described above), it is significant that the Plan retained the simultaneous pursuit of environment and economy as basic ideas.

Additionally, industry has become more aware of the consequences of implementing the Kyoto target. Nevertheless, the norm of economic growth has not returned to prominence, but rather it seems it cannot stand alone without taking the environment into account. For example, industries have committed to voluntary sectoral action plans to reduce GHG, including pursuing further energy efficiency. More positively, some corporations have strategically promoted environmentally sound technologies. Thus, the norm of environmental protection has gradually been absorbed, not only by the public, but also by the corporate culture of industrial sectors. For example, various Japanese companies now prepare "Environment Reports," which are intended to inform stakeholders of their environmental activities.

The integration of all four domestic norms of international contribution, economic growth, energy efficiency, and environmental protection have reshaped the domestic policy in regard to climate change during this stage. The shift is evidenced in the 2005 "Kyoto Protocol Target Achievement Plan," in which the Japanese ability to cope with climate change policies was stated as follows:

> Japan is a resources-poor country in which the foundations of the citizens' lives and industrial activities are dependent on overseas natural resources and which has developed technologies to overcome energy and environmental issues. Moreover, it possesses a

15 The revised objectives of a 6 percent reduction, categorized by gas, are 0.6 percent for energy-originated carbon dioxide; –0.3 percent for carbon dioxide of non-energy-origins; –0.4 percent for methane; –0.5 percent for nitrogen dioxide; 0.1 percent for hydrofluorocarbons, perfluorocarbons, and sulfur hexafluoride; 3.9 percent for carbon sinks; and 1.6 percent under the Kyoto mechanism.

> lifestyle and history in harmony with nature, represented by the concept of "*mottainai*" (literally translated as "don't waste what is valuable"). This adds even more to the reasons why Japan should contribute to the world by presenting a vision of an attractive society which uses natural resources efficiently, making more effort than any other country to achieve the safety and reassurance of the human race and producing results (Government of Japan 2005, 80).

The Japanese were also active internationally in 2005. During that year, Japan participated in two international initiatives. In March 2005, the Ministerial Roundtable on Energy and Environment was organized by the United Kingdom as part of their Group of 8 (G8) process. Twenty countries participated, including the G8, China, India, Brazil, South Africa, and Indonesia. The initiative showed Japan's emerging domestic policy of viewing climate change policies in a more integrated manner, as the roundtable was seen as a test case for integrating the norms of energy efficiency and environmental protection as well as the norm of international contribution by the Japanese.

The other initiative was the Asia-Pacific Partnership on Clean Development and Climate, organized by the United States. The Partnership includes the US, Japan, Australia, China, India, and South Korea (Japan committed to the Kyoto Protocol, but is not satisfied with the situation). Of significance, the initiative included "the major emitters," including the United States, China, and India, which fits well with Japan's struggle to incorporate those countries into the future actions of climate change policies, and demonstrates the relevance of the norm of international contribution that shapes Japanese foreign policy. Japan has struggled to get the United States to join in the international climate change regime, while choosing to endure the current situation of committing to the Kyoto Protocol without the United States. Under the present circumstances the elements of a universal future framework are not clear, while the Partnership created by the United States is seen by Japan as a possible path towards wider participation to combat climate change.

COP11 of the UNFCCC and the first meeting of the Kyoto Protocol (MOP1) were held in Montreal, Canada, during November and December 2005. At COP11/MOP1, Japan insisted that the future framework on climate change policies include global participation with their position being partially supported during the negotiations. COP11 determined that dialogue was necessary for future actions, although it also agreed that the dialogue should not be linked to negotiations. At the same time, COP11/MOP1 initiated the process of future actions in the "second commitment period" for developed member countries on the basis of Articles 3 (9), and the preparation of the review of the Kyoto Protocol based on Article 9. Throughout this process, and based on its earlier framing of climate change, Japan has continued to promote effective climate change agreements with other developed and developing countries. The chapter now turns to reflections on the processes of norm adoption and contestation regarding climate change in Japan.

In summary, after the carnival atmosphere of COP3 when the Japanese eagerly supported action to combat global warming, Japan did not return to the stage where it was overwhelmed by the norm of economic growth. Instead, climate change policy emerged that encompassed the norms of environmental and economic concerns jointly. Domestic policies in regards to climate change policies have shifted from the stage of tackling the global warming problem as one of many global environmental issues, to the stage of addressing global warming as the major social issue, and then to the stage where the problem is treated in a more balanced and complicated manner.

Conclusion

One-twentieth of the world's carbon dioxide emissions in 2000 were from Japan (GESISC 2003, 22). To a certain extent, Japan has responsibility for this serious global environmental problem. As illustrated in this chapter, Japan has accepted its responsibility and has tried to contribute to solving this problem. The question is whether this is sufficient, or furthermore, whether there will be a better path for Japan to take.

Japan's efforts have focused on constructing domestic climate change policies and in developing an international framework on climate change with wide participation. Domestically, climate change policies were developed in line with the Japanese norm of environmental protection, and Japan's climate change policies happened to coincided with its norm of energy efficiency. At the same time, the norm of economic growth has been strong in the policy formulation process in Japan, and gradually the norm of economic growth became strong enough for us to identify it as on equal footing to the norm of environmental protection. Furthermore, the norm of international contribution brought domestic pressure to pursue the strengthening of climate change policies.

Over time, Japan has become proactive and has tried to contribute to the formulation of the international climate change regime based on its norm of international contribution, combined with the norm of environmental protection. These norms, however, also have been influenced by the norms of economic growth and energy efficiency. Japan has promoted international cooperation, first with the United States and Europe, and later it has put more emphasis on assisting developing states to combat climate change. This is evidenced by Japan offering proposals in the international negotiations on UNFCCC and the Kyoto Protocol, and in its current consideration of the configuration of future frameworks on climate change.

As has been demonstrated, existing and emerging norms in Japan conflicted and evolved over time to define and redefine domestic climate change policy. Among these norms, the norm of international contribution after the Cold War remained strong, so that Japan has utilized first global environmental problems and then climate change as an arena for its contribution. Political leadership and public awareness of the issues increased in the periods. Over time, the emerging norm of environmental protection reinforced the norm of international contribution and shaped domestic and international policy. Notably, the norms of economic growth and energy efficiency remain strong in Japan. Yet, they no longer stand alone and dominate as they are now intrinsically linked to the norm of environmental protection. This union started in the early stage but became clearer throughout the second and third stages.

Through the processes illustrated in this chapter, I assert that shifting normative frameworks have shaped and reshaped climate change policies in Japan. In the beginning, Japan failed at the Noordwijk Conference to accept the "Reduction Commitment Norm" but succeeded in committing itself at UNFCCC and further committed to 6 percent reductions of GHG emissions at COP3. In addition, the domestic focus of dealing with global environmental issues produced the Action Plan, which committed Japan to stabilizing carbon dioxide emissions domestically and led to positive contributions to international negotiations for formulating the UNFCCC. Likewise, the normative framework that encompassed the drive to address global warming more directly, fueled Japan's domestic tendency to commit to further reductions with the Joint Conference Report, and in the process the AGBM and COP3. Finally, the normative structure shifted to a more "balanced" one, which considers economic *and* environmental conditions in the post-2012 discussions.

One particular feature of Japan's climate change policy formulation can be identified in a strong inclination toward sharing roles within the government, and among stakeholders. As seen in this chapter, the political leadership of the prime ministers was the strong push necessary to strengthen Japan's domestic climate change policy and its contributions to the international regime. Yet, the prime ministers must overcome inter-ministerial conflicts and promote coordination to generate consolidated policy, and within these efforts all relevant stakeholders needed to be emphasized. This may be perceived as a strength and weakness of the Japanese system. Yet, without these agreements and coordination, Japan could not advance its climate change policy domestically and internationally.

This means that, in the processes of norm contestation, diffusion and adoption, the emission reduction commitment was formulated domestically in the process of inter-ministerial conflict and coordination, and only after the administration committed to the reduction rates (i.e., stabilization of emissions or 6 percent reduction) was Japan able to commit to agreements in the international negotiations. Clearly Japan, as a pluralistic society, has to reconcile various stakeholders' interests in regard to the implementation of climate change policies.

The ministries, as actors, negotiate their individual positions, but are supported by stakeholders in business and civic arenas. As this chapter has demonstrated as well, greater knowledge of the global environment issue has made the Japanese administration formulate domestic programs and actions in more intensive ways. At the same time, other stakeholders, including NGOs and industries, have also gained knowledge. Increasing numbers of environmental NGOs have been established and their activities have enlarged. Industry has shifted to more environmentally sound activities, while the government has become more aware of the consequences of taking action and the need for more comprehensive understanding among stakeholders during the policy process.

Who has the power to determine the commitments? Do agents (Japanese government officials, industry, the media and/or the public) or structures (norms) have the greatest influence? The preliminary answer provided by this chapter seems that no one single actor has created the policy shifts. Clearly, domestic norms have been transformed over time, and in turn, have affected policy directions in Japan. In short, the domestic norms of economic growth, energy efficiency, international contribution, and environmental protection have affected the shaping and reshaping of the Japanese climate change policy.

The question then arises as to why domestic norms on climate change policy have changed in Japan? In some cases, a prominent norm entrepreneur, for example Prime Minister Noboru Takeshita, pushed the process. In others, struggles between competing ministries shaped the union of norms (for example, METI and MoE). Finally, external shifts in international political circumstances (such as the Cold War and post-Cold War), as well as domestic social and economic developments, have affected the power of each domestic norm over time. In tandem, climate change policy has been influenced by the process of Japan's struggle in the world order.

A final question further delineates this issue. Was Japan following or leading international efforts to build the international climate change regime? As the processes of policy development have illustrated, Japan has worked on the development of international schemes concerning climate change from the earliest stages. It has also contributed greatly to international negotiations in formulating the UNFCCC, Kyoto Protocol, and post-2012 schemes. Japan came to place emphasis on the issue of climate change as a policy area that was linked to the fundamental domestic norms of economic growth, energy efficiency, international cooperation, and environmental protection, and was able to translate these domestic norms into international proposals.

References

Akao, N. (1993), *Chikyu wa Uttaeru* (An Agenda for Global Survival). Tokyo, Japan: Sekainougokisya (in Japanese).

Barrett, B. F. D., ed. (2005), *Ecological Modernization and Japan*. New York, NY: Routledge.

Baumert, K. A. ed. (2002), *Building on the Kyoto Protocol: Options for Protecting the Climate*. Washington D. C.: World Resource Institute.

Callon, S. (1995), *Divided Sun: MITI and the Breakdown of Japanese High-tech Industrial Policy, 1975–1993*. Stanford, CA: Stanford University Press.

Checkel, J. T. (1997), International Norms and Domestic Politics: Bridging the Rationalist-Constructivist Divide. *European Journal of International Relations*, 3 (4), pp. 473–495.

Citizens Alliance for Saving the Atmosphere and the Earth (CASA) (1997), *CO_2 Haishutu Sakugen Senryaku no Teigen* (Strategy on CO_2 Emission Reductions). Progress Report (Phase I) (in Japanese).

Finnemore, M. and Sikkink, K. (2001), Taking Stock: The Constructivist Research Program in International Relations and Comparative Politics. *Annual Review of Political Science*, 4, pp. 391–416.

Friedman, D. (1988), *Misunderstood Miracle: Industrial Development and Political Change in Japan.* Ithaca, NY: Cornell University Press.

Global Environment Committee of the Central Environment Council (2004), *Climate Regime Beyond 2012 Basic Considerations.* <http://www.env.go.jp/en/topic/cc/040217.pdf>.

Global Environment Sub-Committee of the Industrial Structure Council (GESISC) (2003), *Perspectives and Actions to Construct a Future Sustainable Framework on Climate Change.* <http://www.meti.go.jp/english/report/downloadfiles/gClimateChange0307e.pdf>.

Government of Japan (2005), *Kyoto Protocol Target Achievement Plan.* <http://www.kantei.go.jp/foreign/policy/kyoto/050428plan_e.pdf>.

Grubb, M. (1999), *The Kyoto Protocol: A Guide and Assessment.* London, UK: Royal Institute of International Affairs.

Hamanaka, H., ed. (2006), *Kyoto Giteisho wo meguru Kokusai Kosho: COP3 Iko no Kosho Keii* (International Negotiations on Kyoto Protocol: After COP3). Tokyo, Japan: Keiogijuku Daigaku Shuppankai (in Japanese).

Hattori, T. (2000), Integrating policies for combating climate change: role of the Japanese Joint Conference for the Kyoto Protocol. *Environmental Economics and Policy Studies*, 3 (4), pp. 25–445.

________. (1999), The road to the Kyoto Conference: an assessment of the Japanese two-dimensional negotiation. *International Negotiation*, 4 (2), pp. 167–195.

Hopf, T. (2002), *Social Construction of International Politics: Identities and Foreign Policies, Moscow, 1955 and 1999.* Ithaca, NY: Cornell University Press.

Huber, T. M. (1994), *Strategic Economy in Japan.* Boulder, CO: Westview.

International Energy Agency (IEA) (2002), Beyond Kyoto: Energy Dynamics and Climate Stabilisation. *OECD/IEA.*

Japan Center for Human Environmental Problems, ed. (1991), *Chikyu Kankyo Mondai to Kokusaiteki Taio* (Global Environmental Problems and International Response). *Environmental Law Journal*, (19: October), (in Japanese), pp. 129–161.

Johnson, C. (1982), *MITI and the Japanese Miracle: The Growth of Industrial Policy, 1925–1975.* Stanford, CA: Stanford University Press.

Joint Conference on Relevant Advisory Councils on Domestic Measures for Addressing Global Warming (1997), *Basic Direction for Measures for the Prevention of Global Warming, Focusing on Comprehensive Measures to Restrain Energy Demand.*

Kameyama, Y. (2003), "Climate change as Japanese foreign policy: from reactive to proactive" in P. Harris, ed. *Global Warming and East Asia.* New York, NY: Routledge, pp. 135–151.

Katzenstein, P.J., R.O. Keohane, and S.D. Kransner, eds. (2002), *Exploration and Contestation in the Study of World Politics.* Cambridge, MA: MIT Press.

Kawashima, Y. (2000), Japan's decision-making about climate change problems: comparative study of decisions in 1990 and in 1997. *Environmental Economics and Policy Studies*, 3 (1), pp. 29–57.

________. (1997a), *Kikohendomondai no Kaiketsu nimuketa Kokusaikyotyo no Kanosei nikansuru Kenkyu* (A Study on the Possibility of International Cooperation towards the Solution of the Climate Change Problem). Dissertation. Tokyo Institute of Technology (in Japanese)

________. (1997b), Comparative Analysis of Decision-making Processes of the Developed Countries towards CO_2 Emissions Reduction Target. *International Environmental Affairs*, 9 (2), pp. 95–126.

Keidanren (Japan Federation of Economic Organizations) (2001), *Request for Calm and Patient Negotiations on the Issues of Global Warming*. <http://www.keidanren.or.jp/english/speech/20010615.html>.

________. (1996), *Keidanren Appeal on the Environment: Declaration on Voluntary Action of Japanese Industry Directed at Conservation of the Global Environment in the 21st Century.* <http://www.keidanren.or.jp/english/policy/pol046.html> (in Japanese).

________. (1991), *Global Environment Charter*. <http://www.keidanren.or.jp/english/speech/spe001/s01001/s01b.html> (in Japanese).

Kiko Forum/Kiko Network (1998), *Kyoto Kaigi karano Syuppatsu* (From Kyoto Forum to Kyoto Network: Start from the Kyoto Conference) (in Japanese).

Mervio, M. (2005), "The environment and Japanese foreign policy: anthropocentric ideologies and changing power relationships" in P. Harris, ed. *Confronting Environmental Change in East and Southeast Asia.* Tokyo, Japan: United Nations University Press, pp. 41–56.

Mintzer, I. M. and J. A. Leonard. eds. (1994), *Negotiating Climate Change: The Inside Story of the Rio Convention*. Cambridge, UK: Cambridge University Press.

Ohta, H. (2005), "Japan and global climate change: the intersection of domestic politics and diplomacy" in P. Harris, ed. *Confronting Environmental Change in East and Southeast Asia.* Tokyo, Japan: United Nations University Press, pp. 57–71.

Paterson, M. (1996), *Global Warming and Global Politics*. New York, NY: Routledge.

Sato, A. (2003), "Knowledge in the global atmospheric policy process: the case of Japan" in P. Harris, ed. *Global Warming and East Asia.* New York, NY: Routledge, pp. 167–186.

Schreurs, M. A. (2002), *Environmental Politics in Japan, Germany, and the United States.* Cambridge, UK: Cambridge University Press.

Schroder, H. (2003), *Climate Change Policy in Japan: From Dusk to Dawn.* Dissertation. Berlin, Germany: Berlin Free University.

Takamura, Y., and Y. Kameyama, eds. (2005), *Chikyuondannkakosyo no Yukue* (Future of Global Warming Negotiations). Tokyo, Japan: Daigakushuppan (in Japanese).

Takeuchi, K. (1998), *Chikyuondanka no Seijigaku* (Politics on Global Warming). Tokyo, Japan: Asahi Shimbunsha (in Japanese).

Tanabe, T. (1999), *Chikyuondanka to Kankyogaiko* (Global Warming and Environmental Diplomacy). Tokyo, Japan: Jijitsushinsya (in Japanese).

The Establishment of Joint Conference on Relevant Advisory Councils on Domestic Measures for Addressing Global Warming. (1997), Prime Minister's Decision (in Japanese).

Wendt, A. (1999), *Social Theory of International Politics*. Cambridge, UK: Cambridge University Press.

Chapter 5

Constructing Progressive Climate Change Norms: The US in the Early 2000s

Cathleen Fogel

By themselves, President George W. Bush's rejection of the Kyoto Protocol in 2001 and his administration's attempts to block any new negotiations on international greenhouse gas (GHG) emission reduction commitments under the Kyoto protocol at the 11th Conference of Parties (COP) in Montreal in 2005, painted a bleak picture of the state of climate change norms in the United States. On second glance, though, things were not as bleak as they seemed. Despite foot dragging in the Presidency, the social diffusion of international climate change policy norms grew steadily in the US in the 2000s, taken up in a range of regional, state, city, corporate and other institutional avenues and dominating national political discourse to an unprecedented degree. Climate change norms had reached such salience by late 2005 that four out of five utility and business leaders predicted that the US would impose mandatory caps on greenhouse gas emissions once President Bush left office (Aston and Helm, 2005). In his 2006 State of the Union Address, President Bush himself declared that "America is addicted to oil," and set a goal of replacing 75 percent of the nation's Middle East oil imports by 2025 with ethanol and other energy sources (Bumiller and Nagourney, 2006).[1]

The premise of this chapter is that climate change norms did spread steadily within the US in the early 2000s, and that these norms included those of binding GHG emission reduction targets and increasing mandatory use of renewable energy.[2] Climate policy entrepreneurs spread these norms in a variety of ways, such as through state legislation and executive orders, city voluntary GHG emission reduction targets, corporate voluntary emission reduction targets, and the adoption

---

1 If implemented, some 15 percent of US petroleum usage might be replaced, as Middle East petroleum imports currently comprise 20 percent of US consumption.

2 The adoption of policies requiring renewable energy are linked to climate change policies in this chapter due to the primacy of three justifications for policies promoting renewable energy: (1) to decrease greenhouse gas emissions from electricity production that would otherwise originate from fossil fuel sources; (2) to decrease the emission of local air pollutants that cause health problems such as asthma; and (3) energy security. Of these three rationales, the first was most commonly used during the period discussed in this chapter.

of voluntary targets by churches. The story that this chapter presents is that climate policy entrepreneurs opportunistically advanced a variety of normative frames to promote and justify climate action based on their target audience and fora at any given moment. In other words, despite President George W. Bush's 2001 declaration that "the Kyoto Protocol is dead," norms embodied within the protocol continued to spread and become imbued with a legal character as they were enacted into law at the state level. Contextual changes, such as the faltering US war in Iraq, rising oil prices and the devastation wrought by hurricane Katrina greatly accelerated and shaped the diffusion of these norms, as did the entry into force of the Kyoto Protocol and the European Union Emissions Trading Scheme in 2005, and new, high-visibility green branding campaigns by US corporate giants General Electric, General Motors and Walmart in 2005 and 2006. In sum, between 2001 and 2006, a discernable shift on climate change occurred in the US, reframing public consciousness and drawing a line between those entities engaged in climate change solutions and those lagging behind.

This chapter explores the nature of the domestic climate change and renewable energy norms and policy developments that occurred between 2001, when George W. Bush became President, and 2006, when former Presidential contender Al Gore's climate change movie, "An Inconvenient Truth," hit theaters nationwide. What was the nature of the GHG emission reduction commitments adopted by state legislatures and city capitols across the nation? What resemblance did these bear to Kyoto Protocol emission reduction norms, and what influence did these norms bring to unfolding US national climate policies? Did climate innovations occur only in "blue" coastal Democratic states, or were advances relatively evenly spread between these, and more conservative and Republican "red" central, southern and mountain states? What other leading domestic policy actors adopted and/or worked to diffuse GHG emission reduction and renewable energy norms?[3] Did developments, on the whole, amount to a "tipping point" beyond which significant federal climate policy action became inevitable or were they relatively reversible? Finally, what narrative framings or storylines (Hajer 1995) did climate policy entrepreneurs advance in the 2000s, and what significance did these have for the engagement of new and powerful domestic actors and coalitions?

This chapter draws theoretical resources from work on international norms, discourse coalitions and cultural values. Loren Cass, in his chapter in this volume, reviews in detail the importance of norms in international relations, be they "constraining" or "constitutive" in nature. In the first conception, regulative norms constrain actors' behavior by altering the incentive structures that they face; in the latter, norms affect behavior through learning processes in which the norm becomes fundamental to actors' identities and interests (These two conceptions are also present in the Pettenger and Hattori chapters). The distinction is relevant in that regulative norms are frequently adopted via coercive or "shaming" processes, while

3 This chapter therefore explores relevant agents in causing national policy change, as does the chapter by Hattori, also in this volume, but with significantly different results.

constitutive norms are generally spread through persuasion and the internalization of new values and understanding.

Maarten Hajer's work presents a framework for understanding how norms, expressed through talk or discourse, are translated into new environmental institutions and policies. Discourses are ensembles of "ideas, concepts and categorizations that are produced, reproduced and transformed in a particular set of practices and through which meaning is given to physical and social realities" (Hajer 1995, 44). They frame certain problems and ignore others and typically contain both norms and analyses. By forming the context in which phenomena are understood, discourses prefigure definitions (and solutions) to problems (Hajer 1995, 45). A process called "discourse institutionalization" occurs as discourses solidify into institutions through policies, organizational practices or dominant ways of reasoning. Realist accounts of international relations typically ignore domestic policy discourses, while constructivists view them as integral to understanding.

Discourses may also be conceived of as storylines, generative narratives which serve as the mediums through which actors seek to impose their view of reality on others. "Story-lines are devices through which actors are positioned or framed, and through which specific ideas of 'blame' and 'responsibility,' and of 'urgency' and 'responsible behavior' are attributed" (Hajer 1995, 65). "Discourse coalitions" are the ensembles of a set of storylines and frames, the actors that utter these, and the practices that conform to these storylines, all organized around a discourse.

Factors influencing actors' acceptance of emerging storylines include the attractiveness (or apparent necessity) of the subject-positioning in which a storyline places listening actors; the credibility of the positioning and the overall story structure to listening actors; and the trust actors place in either institutions or persons creating the storyline (Hajer 1995). If actors dislike the position in which they have been narratively framed, or if they distrust those uttering the storylines, they tend to reject it. Discourse coalitions tend to form when "previously independent practices are being actively related to one another, and a common discourse is created in which several practices get a meaning in a common political project" (Hajer 1995, 65).[4]

Thomas Rochon (1998) takes these ideas one step further by outlining the ways in which the values or norms embedded within storylines and discourse coalitions are formed. Values are shaped and diffused throughout society in three different ways: "Value conversion" involves fundamental reconceptualizations of previously existing categories or beliefs; "value creation" involves the creation of new terms in situations where the phenomena was not previously taken into account; and, "value connection" involves forging conceptual links between phenomena previously thought to be unrelated. The latter two (value creation and value connection), Rochon believes, are the easiest ways to convert people to new ways of thinking.

This chapter uses these theoretical tools to explore the diffusion of progressive climate change norms throughout the US in the early 2000s. It defines "progressive

---

4 See also Bäckstrand and Lövbrand, this volume, for further discussion of discourses in climate change politics.

climate change norms" as policy or behavioral changes to achieve greenhouse gas emission reduction targets that meet or exceed those contained in the Kyoto Protocol and/or that endorse long-term emission reduction targets of up to 60–80 percent by 2050. The chapter begins with a description and analysis of various, important US domestic policy actors and institutions, and the discursive framings and norms that they adopted. What we find is an increasing variety of progressive climate change norms becoming institutionalized in sub-national institutions. In turn, these newly legitimated norms resulted in further expansion of domestic climate change discourse coalitions to new actors and new institutions in a formidable cycle of change.

States

The most heartening and important climate change policy development in the early 2000s was the groundswell of state-level action on climate change and renewable energy policy. Some of this action grew out of state-level statements and policies on energy and climate dating to the late 1980s, when climate change concerns first burst onto the international political stage and into domestic political consciousness. But other critical developments, such as the 2001 adoption in California of strict automobile GHG emission tailpipe standards, appeared seemingly out of nowhere, reflecting climate entrepreneurs' frustration with and increasing creativity in overcoming federal inaction. In all cases, the benefits associated with GHG emission reductions (such as decreasing air pollution, cutting costs through energy efficiency, and in particular, concerns about the economic costs of climate impacts) were vital to the measures' acceptance. "States that enact climate change policy almost always do so with long-term economic well being in mind" (Rabe 2002).

By early 2006, 21 states and the District of Columbia had mandated that a certain percentage of electricity generation originate from renewable sources. Referred to as "renewable portfolio standards" (RPS), the targets for state energy supply from renewable sources varied from the ambitious (with California law mandating that 20 percent of electricity from included utility generators come from renewable sources by 2017) to the modest (Wisconsin, which calls for 2.2 percent of electricity from renewable sources by 2011). Justifications for action similarly ranged from energy security in Texas, to job creation and competitiveness in California, to improved air quality in many locations. Although not explicitly identified as climate change policies, electricity generation accounted for some 30 percent of overall GHG emissions in the US in 2005, therefore, increases in renewable energy generation at the state (and federal) level were rightly seen by climate policy entrepreneurs as holding great promise for decreases in national GHG emissions.

States also aggressively developed public funds, consumer choice and technical standards to increase the proportion of their energy supplies that originate from renewable sources. Over 20 states had, or were developing, public funds to support renewable energy and/or energy efficiency projects in their states via fees added to consumers' electricity bills. Thirty-six states offered green pricing, in which consumers could choose to support renewable energy, and five states had made green pricing programs mandatory. Some 41 states offered "net metering" options to allow corporations and individuals to sell electricity generated on-site via renewable sources back to the grid (Rabe 2002). State renewable energy promotion programs were predicted to increase energy generated from renewable sources from about 12,000 megawatts (MW) in 2006 to 34,000 MW in 2017 (UCS 2005).

State renewable energy programs revealed the increasingly bi-partisan nature of climate change-related policy action and diverse motivations. In Texas, for instance, promotion of renewable energy had been based, in part, on focus groups that identified citizen concern about energy supply reliability for the state, itself stemming from Texas' status as a net energy importer in the 1990s. In 1999, Texas created a RPS through an electrical utility restructuring bill, which was signed into law by then-Governor George W. Bush. By 2002, the "Texas Wind Rush" had increased operational MW of wind power generated in the state from 187 MW to 1,101 MW (Rabe 2002).

In June 2005 in California, Republican Governor Arnold Schwarzenegger called for advancing the proportion of the state's energy supply generated from renewable energy by 2017 from 20 percent to 33 percent. Accompanied by ambitious new GHG emission reduction targets for the state, this higher state RPS target was justified both by an explicit acceptance of climate science and by predictions of job increases and improvements in California's competitiveness in renewable energy technologies in the face of recent German and Japanese gains.

Mr. Schwarzenegger called for California to reduce its GHG emissions to 2000 levels by 2010, to 1990 levels by 2020, and to 80 percent below 1990 levels by 2050—the most ambitious long-term targets yet announced in the US. On June 9 of 2005, New Mexico Governor Bill Richardson set similar GHG reduction targets, the first ever for a major coal, oil and gas producing state. This brought to nine the number of states with GHG emission reduction targets (California, New Mexico, New Jersey, Maine, Massachusetts, Connecticut, New York, Washington and Oregon) as of early 2006.

In addition, some 28 states had produced Climate Action Plans by 2006, with programs ranging from energy efficiency and demand management, to forestry and carbon sequestration, waste reduction, transportation policies, building design and industrial energy co-generation (Pew Center 2005). Most states benefited from technical and financial assistance for these from the federal Environmental Protection Agency's (EPA) State and Local Climate Change Outreach program, just one of many federal programs that had quietly built capacity and continued to support climate policy innovation at the state level during the early 2000s. With EPA assistance, 41 US states had initiated or completed GHG emission inventories (Anonymous 2002). In states such as Texas, California and New York, GHG emissions exceeded those of the United Kingdom, France and the Netherlands, respectively, so action at the US sub-state level was not insignificant. In 2005, state environmental legislators were represented for the first time at Kyoto Protocol negotiations by the newly formed National Caucus of Environmental Legislators.

In the area of transportation, California school teacher-turned-legislator Fran Pavley, led passage of a tough new bill in 2002 requiring automakers selling in the state to reduce tailpipe emissions of CO_2 and other pollutants by 22 percent by 2012 and 36 percent by 2016. Signed into law that same year by then-Governor Davis, a coalition of automobile makers in 2005 sued to prevent the regulations from entering into force. Nevertheless, by 2006, ten states had announced their intention to abide by the California standards (New York, Maine, New Jersey, Vermont, Massachusetts, Oregon, Washington, Rhode Island, Connecticut and Pennsylvania).[5] Passage of the California bill, and state endorsements of it, rested in large measure on concerns with climate change's expected impacts in California, including alterations in precipitation and temperature patterns, rising sea levels, and increased ground level ozone pollution (CARB 2006).

In addition, several states began experimenting with carbon offsets and sequestration in the early 2000s. Washington and Oregon required new power plants to neutralize anticipated CO_2 emissions through emission reduction, mitigation or offset projects. The states of Massachusetts and New Hampshire required GHG emission reduction actions from existing power plants. In California, the state's largest utility announced a voluntary carbon offset program, whereby consumers could elect to participate through fees that then subsidize reforestation and forest conservation within the state (PG&E 2006).

In the Midwest, the agricultural states of Illinois, Nebraska, Oklahoma, North Dakota and Wyoming formed carbon sequestration advisory committees in the late 1990s in order to explore opportunities for sequestration in the farming sector. The state of Illinois acted on the findings of its Committee in early 2006 with the initiation of the Illinois Conservation and Climate Initiative (ICCI), a voluntary program that would award carbon credits to farmers for GHG reduction practices (including farm-land conservation tillage, tree and grass planting, and methane capture). Credits produced in the program were to be sold to the Chicago Climate Exchange (CCX) following their certification by an independent third-party auditor (Pew Center 2006). Offset programs such as these benefited from framings emphasizing the economic opportunities inherent in climate change mitigation, as well as predicted climate impacts, and economic need in the reality of limited or declining state budgets.

The policy actors that motivated these changes at the state level often operated at the national level. For instance, the non-governmental organizations the Union of Concerned Scientists and the Climate Action Network's State Policy Advocates Network worked nationally during this period. These types of groups received research and funding support for state level change from foundations such as the Energy Foundation and the Pew Center. Additionally, state and local policy actors

5 The federal Clean Air Act allows states to set limits on tailpipe emissions and allocates regulation of automobile fuel efficiency standards to federal government. In late 2005, the Bush Administration issued weak fuel-efficiency regulations that, amongst other things, included a federal preemption against state laws attempting to reduce CO_2 emissions from cars.

and organizations closely tracked and influenced the above progressive climate change gains.

Regions

The diffusion of international climate change norms within the US in the early 2000s also became evident in regional climate change accords, such as those resulting from the Conference of New England Governors and Eastern Canadian Premiers (NEG/ECP) and a West Coast Governors' Global Warming Initiative (WCGGWI). Building in many cases on previous collaborations, regional initiatives presented the greatest opportunities for the rapid diffusion of progressive climate change policy norms in the US in the early 2000s.[6]

In 1988, the NEG/ECP adopted regional plans to reduce acid rain and limit mercury pollution. In 2001, building on the trust and shared concerns so developed, the NEG/ECP released a Climate Action Plan that committed signatory states to GHG emissions reductions to 1990 levels by 2010 and, over the long term, to levels sufficient to avoid dangerous climate change (interpreted later as cuts of approximately 75 to 85 percent below 2001 levels by 2050) (NEG/ECP 2001). Six New England states signed the NEG/ECP plan.

In early 2003, Governor George Pataki of New York proposed GHG reduction targets to implement the accord in his state, as well as a Regional Greenhouse Gas Initiative (RGGI) that would coordinate a regional system to cap and trade carbon dioxide emissions from major power plants. Over the next two years, eleven New England states participated in designing the RGGI program, the details of which were announced with support from seven Governors in late 2005 (New York, Connecticut, Delaware, Maine, New Hampshire, New Jersey and Vermont).[7] The initiative set the goal of maintaining current CO_2 emissions levels of approximately 120 million metric tons of CO_2 between 2009 and 2015; emission levels would then be reduced by 10 percent by 2019. Included in the initiative were about 180 power plants of at least twenty five megawatts that derived at least half of their energy from fossil fuels. Power plants were permitted to meet their CO_2 targets via direct emission reductions (e.g. efficiency measures, fuel switching), buying allowances, or the purchase of a limited number of emission offset credits for projects outside of the power sector such as landfill methane, afforestation, and heating oil efficiency.[8] To stimulate further development, renewable energy and energy efficiency programs

---

6 Foundations prioritizing work to build regional GHG initiatives include the Emily Hall Tremaine Foundation, the Energy Foundation and the Pew Charitable Trusts. Primarily policy actors ranged from governmental staff members and politicians to local non-governmental organizations and state/federal NGOs.

7 Representatives from the states of Massachusetts and Rhode Island participated in RGGI's design, while the states of Maryland and Pennsylvania participated as observers.

8 Power generators would be limited to covering 3.5% of their emissions with offset allowances, an amount equal to about half of a source's emission reduction requirement.

were to receive about one quarter of the program's initial emission allocations. A Regional Greenhouse Gas Registry (RGGR) was created to accurately track emission reductions and trades. Both the RGGR and the RGGI itself were designed so that additional states could join them or form compatible sister programs at any time (Pew Center 2005).[9]

One such program developing in 2006 was the WCGGWI. Initiated in 2003, Governors from the states of Oregon, Washington and California formed the regional initiative "individually and regionally to reduce greenhouse gas emissions" after concluding that global warming would negatively affect the health and economies of their states (WCGGWI 2003). They launched five working groups (in the areas of emission reporting, energy efficiency, renewable energy, ports and highway diesel emissions and hybrid procurement) and directed their staff to develop joint policy proposals for regional cooperation. Staff recommendations in these areas were announced in late 2004. In addition to proposals to spur hybrid vehicle purchases, as well as energy efficiency and renewable generation measures, the recommendations highlighted the importance of adopting regional and state GHG emission reduction targets, shared vehicle emission standards and carbon allowance (or cap-and-trade) programs (WCGGWI 2004).

Parallel to this, the California Energy Commission in the early 2000s began coordinating eleven Western states in the development of a voluntary Western Renewable Energy Generation Information System (WREGIS). This tracking system was to be used by participating states to track the generation and trade of renewable energy credits (RECs) generated in part to meet state-level renewable energy portfolio standards. Many of these Western states also participated in the Western Governors' Association's Clean and Diversified Energy Initiative, which explored energy efficiency and renewable energy opportunities for an even larger eighteen state Western region (WGA 2004). Similarly, elected and government officials from five Midwest and Northern Plains states began a Powering the Plains effort in 2001 in order to study and encourage job creation by promoting biofuels, wind power and other sources of alternative energy. They presented a broad plan for renewable energy production in June 2006 (Kehrl 2005).

As with state programs, the New England RGGI initiative had been framed by proponents as having the potential to decrease energy costs, benefiting both consumers and state economies. The non-profit Natural Resources Defense Council (NRDC) argued that RGGI could, "save the average household between $30 and $50 a year." As RGGI moved closer to implementation, however, these claims came under dispute. Massachusetts Governor Mitt Romney withdrew from the pact in late 2005 along with the Governor of Rhode Island; Romney argued that implementing the RGGI might increase consumer energy bills by 10 percent, a reversal of his previous position (Griscom Little 2005). Instead, Romney introduced a state plan that would set a price cap for traded GHG emissions credits (Griscom Little 2006). This reversal followed a lobbying effort by the Associated Industries of Massachusetts

9 Eastern Canadian Provinces may also join the RGGI initiative.

(AIM) earlier that year, in which AIM framed the regional initiative as likely to lead to increased energy costs, economic-competitiveness problems, job losses, increased health care costs and costs to local governments.

As this vignette indicates, not all state-level leaders supported climate change or action on renewable energy action in the US in the early 2000s. One particularly effective group working against emerging progressive climate change norms was the American Legislative Exchange Council (ALEC). Backed by conservative corporate sponsorship, ALEC's members in 2005 included some 98 members of Congress, more than 100 state legislators and five governors (Marzilli 2005). In 2004, ALEC distributed model "Sons-of-Kyoto" legislation, calling on states to reject "every form of Kyoto legislation," because it is "just another highly regressive energy tax on America's working families."[10] As in the Massachusetts case, national ALEC leaders argued that reducing GHG emissions under the Kyoto pact would increase electricity and natural gas prices by 86 and 147 percent respectively, and cost the nation between $130 and $400 billion annually. Discounting threats from climate change, ALEC leaders argued that "carbon dioxide … is beneficial to plant and human life alike" (ALEC 2004).

As of 2002, 16 states had passed legislation or resolutions criticizing the Kyoto Protocol and urging the US Senate to refuse to ratify the pact.[11] Michigan passed legislation prohibiting state agencies from any work related to reducing GHG emissions estimates or reductions unless requested to do so by the legislature. West Virginia adopted a similar law, and neither state has pursued the issue in the early 2000s. ALEC also worked with members of the US congress to pass federal preemptions of state environmental laws that it opposes, such as the state of California's regulation of automobile GHG tailpipe emissions.

While the discursive battle over the Kyoto Protocol and the norms that it represented (GHG emissions reductions targets and expanded use of renewable energy) continued to rage in 2006, it seemed increasingly unlikely that norms newly embodied in progressive state and regional laws and programs would be reversed, in part because an increasing number of states framed climate change in terms of the likely economic costs of inaction and the potential economic opportunities of taking the lead. Opponents challenging this frame continued to stress the disputed and increasingly dated frame of the economic burden of climate action.

---

10 The ALEC model legislation "prohibits the proposal or promulgation of state regulations intended to reduce emissions of greenhouse gases, prior to ratification of the Kyoto climate change protocol by the United States senate and the enactment of implementing legislation by the United States Congress" (ALEC 2006).

11 These states are: Alabama, Arizona, Colorado, Idaho, Illinois, Indiana, Kentucky, Michigan, Mississippi, North Dakota, Ohio, Pennsylvania, South Carolina, Virginia, West Virginia, and Wyoming (Rabe 2002).

Cities, Universities, Schools, Churches and Unions

Many cities, universities, schools, churches and unions increasingly articulated progressive climate protection discourses in the 2000s, alongside states and regions. Similarly to states, city and county governments do not have authority outside their own jurisdictions, and budget limitations restrict the scope of their activities. Nevertheless, trends in the 2000s indicated a growing diffusion of climate norms and policies amongst US local domestic leaders and institutions, including serious GHG emission reduction commitments.

One of the most far-reaching programs was that of the Cities for Climate Protection (CCP) Program of the International Council on Local Environmental Initiatives (ICLEI). As of 2006, 159 US cities had joined this effort. Participation in the CCP involves a five-step program, including assessing GHG emissions, setting reduction targets, planning, implementation, and monitoring and verification (ICLEI 2006). As might be expected, West and East coast cities dominated the CCP program, but unexpected leaders emerged in the "red states" of Utah, with Salt Lake City Mayor Rocky Anderson, and Illinois, with Mayor Richard Daly of Chicago (Marzilli 2005). GHG emission reduction commitments and the program steps of the CCP mirrored those at the state level and included waste reduction, promotion of renewable building, energy efficiency, building design, tree planting and green fleet programs, among other programs (ICLEI 2006).[12]

In a similar vein in 2005, Seattle Mayor Greg Nickels led the US Conference of Mayors in its endorsement of a "US Mayors Climate Protection Program," signed by 200 mayors from 38 US states. Cumulatively, these 200 mayors represented 41 million citizens and their numbers included those from states typically resistant towards, or opposed to, climate protection policies, such as Kentucky, Georgia, Missouri, North Carolina, Florida, Michigan and Montana.[13] The Mayors statement called on federal and state governments to enact Kyoto Protocol-like reduction targets (7 percent below 1990 levels by 2012) and committed local government signatories to "strive to meet or exceed Kyoto Protocol targets for reduction global warming pollution by taking actions in our own operations and communities" (Seattle gov. 2005). In September 2005, ICLEI announced the linked target of reducing 20 million tons of CO_2 equivalent emissions by the close of 2007, pledging that it would work with the 200 mayors and cities to accomplish this (ICLEI 2005). This was a realistic target, it seemed. In 2004, participants in the ICLEI program reported approximately 23 million tons of CO_2 equivalent reductions, estimating that this saved cities almost $725 million in reduced energy and fuel costs.

In 2006, the City of Austin, Texas launched an innovative program called the "Plug-in Partners" campaign. In the program, Austin aimed to enlist 50 US cities

---

12 For a detailed study of the adoption of climate change policies by cities see Bulkely and Betsill (2004). See also the Sierra Club for information on its "Cool Cities" program <http://www.coolcities.us/>.

13 See map at: <http://www.ci.seattle.wa.us/mayor/climate/default.htm#who>.

in creating market demand for Plug-in Hybrid Electric Vehicles (PHEVs). To launch the campaign, city officials placed 'soft orders' for 600 of PHEVs and set aside $1 million for citizen rebates. The city claimed that "electric gas" cost less than $1/gallon and that PHEV's were rechargeable overnight during hours of surplus electricity generation while further noting that PHEVs could achieve 100 miles/per gallon of gasoline (Austin Energy, 2006).

US universities caught the climate protection bug in the 2000s as well. In the Northeast, the Campuses for Climate Action project of the non-profit Clean Air–Cool Planet began to provide GHG emissions inventory advice and encouraged universities and colleges to assess and reduce their GHG emissions (Clean Air Cool Planet 2006). Nationwide, more than 110 universities and colleges in 2006 had built or were building structures certified by the US Green Building Council, an organization promoting building constructions and designs with high standards of energy efficiency and energy conservation (Egan 2006). In addition, 95 New England colleges and universities had formally committed to supporting specific GHG reduction targets while the University of California launched a similarly ambitious program (Tufts Climate Initiative 2005).

A number of religious associations sprang into action on climate change in the early 2000s as well. Perhaps most inspiring was the Interfaith Power and Light project, founded by Reverend Sally Bingham in 2001 (Interfaith Power 2006). Operating out of the Episcopalian Grace Cathedral in San Francisco, the Interfaith Power and Light project aimed to educate congregations about the theological basis for eco-stewardship; it also provided tools and information about reducing greenhouse gases. By 2005, the project had worked with more than 300 California congregations to conserve energy and to prevent over 40 million pounds of carbon dioxide from entering the atmosphere (Pope 2005). Interfaith Power aims eventually to inspire change in all of California's 50,000 religious congregations (Interfaith Power 2006). In 2006, the Interfaith Power and Light project distributed copies of the movie "Inconvenient Truth," narrated by Al Gore, to some 4,000 religious congregations throughout the US, raising awareness and stimulating debate about climate change, and moral and practical responses to it.[14]

Other compelling initiatives from the US religious community included the 2003 "What Would Jesus Drive?" Road Tour, in which the Reverend Jim Ball, Executive Director of the Evangelical Environmental Network (EEN), toured eight states and eleven cities touting transportation choices as moral choices. "The Campaign believes that Jesus wants people to drive the most fuel-efficient, least polluting vehicle that truly meets their needs" he said (EEN 2006). Ball's tour spawned a flurry of debate and media coverage, as well as a dozen state-level church outreach campaigns promoting fuel efficiency and GHG emission reductions in "what should the governor drive?" campaigns (ICCN 2006). Formed because it "recognize(s) [that] many 'environmental' problems are fundamentally spiritual problems," EEN's

---

14 See <http://www.interfaithpower.org/index.htm> for more information on this outreach campaign.

outreach efforts included publishing a Creation Care Magazine, hosting talks by climate scientist Sir John Houghton, a "devout Christian," and producing a hard-hitting "Global Warming Briefing for Evangelical Leaders." The latter infused a religious community storyline into climate discourses by its simple statement that, "addressing global warming is a new way to love our neighbors" (EEN 2006).

Other religious leaders' initiatives to prompt action on climate change were supported and reported on by the National Religious Partnership for the Environment (NRPE). In 2002, the Interfaith Climate Change Network (ICCN) of the NRPE sent a "Religious Leaders Statement on Energy Conservation" to US Senators, signed by 1,200 religious leaders. In a collaborative effort of the Eco-Justice Working Group of the National Council of the Churches of Christ in the USA and the Coalition on the Environment and Jewish Life, the ICCN (in campaigns in 21 states) organized and presented similar statements that same year to the CEO's of Ford, General Motors and Chrysler, urging action on automobile fuel economy standards. The previous year, the US Conference of Catholic Bishops unanimously backed a statement calling for immediate action to mitigate climate change; and in 2004, the Conference wrote US Senators urging immediate climate change action (USCCB 2004). The National Council of Churches, an ecumenical partnership of Christian churches representing some 100,000 congregations, described stewardship of the earth as a "moral value" (Pope 2005). In February of 2006, these efforts were capped with the release from 86 Evangelical leaders of a statement calling for federal action on climate change because "millions of people could die in this century because of climate change, most of them our poorest global neighbors." The statement represented the first stage of an ongoing "Evangelical Climate Initiative" slated to continue throughout 2006–2007 with television and radio spots in states with influential legislators, informational campaigns in churches, and educational events at Christian colleges (Goodstein 2006). In sum, moral framings on climate change and energy conservation, including actions to secure GHG emission reductions, staked out a significant discursive space in US religious communities' discourse in the early 2000s to the point of nearly becoming "mainstream."

In the labor sector, a new organization called the "Apollo Alliance" was launched in 2000 with the aim of reframing energy and climate issues as national security and economic issues. The goal of the Alliance was to mobilize federal commitment to energy independence; and major national labor unions (such as the United Steelworkers of America) were represented on the Apollo Alliance Advisory Board. Environmental organizations and prominent business leaders were also represented and perhaps not surprisingly, the organization endured bitter disputes over specific climate and energy-related legislation. All of the Alliance's members, however (including the labor unions) endorsed its broad goals: a federal renewable portfolio standard that would require 15 percent of US electricity to come from clean energy sources by 2015; a renewable fuel standard, that would require 10 percent of fuels to be derived from biomatter by 2010; and strong efficiency codes for buildings. A federal renewable fuel standard was, indeed, adopted in 2005.

As of 2006, however, the Apollo Alliance had not yet taken a position on key climate change federal policies, such as proposed increases corporate fuel economy (CAFÉ) standards (that increase automobiles mileage per gallon), gasoline taxes or mandatory caps on carbon dioxide emissions. The Alliance did however call for $300 billion in federal clean energy subsidies over ten years. Mixing economic opportunity storylines with climate protection, the Alliance supported this platform with predictions that such subsidies could produce over three million new jobs and reduce national energy consumption by 16 percent. This framing reflected a major rethinking of climate narratives that occurred throughout the US following the release of a controversial report in 2004: "The Death of Environmentalism." This report excoriated US environmental organizations for their outdated "crisis" framing of climate change and called instead for a reframing in terms of the economic and job opportunities of climate protection (Griscom Little 2005; Shellenberger and Nordhaus 2004). This shift matches the discourses surrounding climate change identified by Paterson and Stripple in this volume.

Labor unions likewise joined the National Commission on Energy, along with environmental, business and former government leaders. The Commission argued for price-caps on federal GHG emissions reductions, rather than non-negotiable emissions caps. While controversial amongst environmental leaders, this proposal nonetheless signified the entry of previously staunchly opposed labor groups into climate change discourses on carbon caps (Griscom Little 2005a). Like discursive changes reflected amongst city, university and religious leaders, this framing reflected the increasingly broad swath of American society concerned about climate change that worked for concrete governmental action to address the problem.

Corporations, Investors, and Insurers

Many corporations and financial institutions took up climate change discourses in the early 2000s, and in some cases this included commitment to, and concrete action on, GHG emission reduction as well. Some corporate action was prompted by public shaming campaigns such as that mounted by the non-governmental organization Rainforest Action Network's (RAN) global finance campaign. RAN campaigned against Citigroup from 2001 to 2003, until that institution adopted a corporate environmental and climate change policy. In the succeeding years, many additional US banks followed suit.[15] Corporations also joined market-based institutions such as the CCX, voluntarily pledging to reduce GHG emissions and then trading carbon credits in order to do so (CCX 2006). California businesses began to participate in voluntary GHG emissions reporting efforts, such as that of the California Climate

15 Targets of the RAN global finance campaign have included Citigroup, Bank of America, JP Morgan Chase, Wells Fargo, and most recently, Canadian Banks involved in financing the expansion of oil extraction from tar sands in the Canadian province of Ontario (RAN 2006).

Action Registry, and international finance giant HSBC bank broke new territory in climate norms and green branding by pledging in 2005 to become "carbon neutral."

The best known corporate climate engagement program in the US in the early 2000s was the EPA's Climate Leader Program. With Citigroup's January 2006 pledge to reduce its GHG emissions footprint by 10 percent by 2011, 46 corporations in total had by then committed to GHG emissions reductions as part of their participation in the EPA Climate Leader Program. Eighty companies participated in the EPA program, cumulatively emitting together about 8 percent of total national GHG emissions (EPA 2006b). Those that made GHG emission reduction commitments included household names such as Frito Lay, the Gap, 3M, Dupont, Starbucks, General Motors, Caterpillar, Johnson & Johnson, Eastman Kodak, Bank of America, Pfizer, General Electric and Hewlett Packard. In addition, a new international public-private network, The Climate Group, was founded in 2004 with the aim of documenting and then publicizing the economic opportunities associated with corporate GHG emission reductions; this reframing effort was highly successful as reflected in an increasing number of well-placed media stories and an increasing permeation of the economic opportunity frame into corporate advertising and reporting (EPA 2006a; The Climate Group 2006; Aston and Helm 2006). Understandings of climate change as presenting economic opportunities were catapulted forward when General Electric Corporation announced its "EcoImagination" program to ramp up clean energy production in late 2005 (Barringer and Wald 2005; Ignatius 2005).

In 2002, climate policy entrepreneurs launched the Carbon Disclosure Project, an effort to elicit climate related information from Fortune 500 companies on behalf of one hundred public and private investors. By 2005 the project had elicited climate change and energy information from 71 percent of Fortune 500 corporations; 90 percent of which identified climate change as posing commercial risks while half disclosed GHG emissions data and reported having emissions reductions programs and/or targets. About 6 percent of the corporations surveyed reported GHG emission reductions (CDP 2005).

In December 2005, movements simmering in insurers and state public fund circles came to a boil. In that month, 20 large US investors wrote to 30 of the largest publicly-held US insurance companies, requesting that they disclose their financial exposure to climate change as well as the steps they were taking to reduce financial impacts. Participants in the Investors Network on Climate Risk (INCR), which organized the letter, included state treasurers and controllers from eight states, two of the largest public pension funds in the US (CalPERS and CalSTRS), the New York City Comptroller, the Illinois State Board of Investment, socially responsible investment funds, faith-based investors, and labor pension funds. INCR members collectively controlled some $800 billion in assets at that time (Anonymous 2006). The investors' concerns corresponded to those of a 2005 study that had documented a 15-fold increase in insurance losses from catastrophic weather events over the preceding three decades in the US, warning of further financial loses from climate change in the future (CERES 2005).

Federal Legislation and Presidential Candidates

Of course, state, corporate and city discourse and action on climate change (however inspiring) were by themselves no substitute for coherent national climate change policies, GHG emission reduction targets or renewable energy laws. Nor did they substitute for US participation in the Kyoto Protocol. The discourse and GHG emission reduction pledges and other developments detailed so far, however, did appear to be forming the foundation for a deep shift in US climate change discourses and policy norms. What were the chances that federal actors or the US Congress would soon embrace aggressive action to reduce GHG emissions and promote renewable energy?

In 2005, after years of debate, a US national energy bill was signed into law by President George W. Bush. Opposition from conservative legislators had resulted in the removal of many of the most far-reaching climate change actions from the bill. The bulk of the bill's $14.5 billion in subsidies went to the oil, coal, gas and nuclear industries, while only $4.3 billion in renewable energy, energy efficiency and conservation subsidies were adopted. However, the provisions of this bill would double national ethanol production by 2012. In the context of the US in 2005, this represented progress. Nevertheless, stripped from the 2005 energy bill were provisions to reduce oil consumption by a million barrels a day by 2015 as well as a federal renewable electricity portfolio standard of 10 percent by 2020 (Griscom Little 2005a). In separate action, new corporate fuel efficiency rules announced by the Bush Administration in 2005 increased standards for passenger cars to a mere 27.5 mpg and that of SUVs to 24 mpg, a move widely derided as ineffectual.

Federal climate policy discourse nonetheless swerved in the 2000s. It began to reflect the broad movements in American society discussed previously, but also the more immediate public concerns about rising oil prices, the US presence in the Middle East, and the horrors of Hurricane Katrina. In 2003, Senators John McCain (R-AZ) and Joe Lieberman's (D-CN) had introduced the Climate Stewardship Act, which would have capped national GHG emissions. The bill failed, but by a surprisingly close 43–55 vote. In 2005, presidential-hopeful McCain reintroduced the proposal as an amendment to the federal energy bill, adding in the process a host of "pork barrel" incentives to the nuclear power industry. In addition, President Bush also argued about the appropriate technologies to solve climate change in 2006, saying "nuclear power will help us deal with the issue of greenhouse gases" (Rutenberg 2006). Discourse such as this on the ability of various future technological advancements to solve climate change began, in 2005 and 2006, to reshape climate change as a "technology issue." Booming sales of the hybrid Toyota Prius in 2005 and the increasing organization of the nuclear lobby to capitalize on climate debates helped trigger this development.

Another likely 2008 Republican presidential contender, Senator Chuck Hagel (R-NE), introduced three climate and energy bills in 2005. These proposed a range of voluntary public-private partnerships and corporate incentives for clean energy technologies. Senator Jeff Bingaman (D-NM), a third aspirant to the presidency, introduced a federal GHG emissions reduction bill that would cap prices of carbon at $7/ton, rather than cap carbon quantities. Again, proposing penalties on carbon emissions, although not a call for an inflexible cap on GHG emissions, represented a discursive breakthrough of sorts.

Other presidential hopefuls, such as Mitt Romney, Governor of Massachusetts, and George Pataki, Governor of New York, participated in or led the Northeastern RGGI initiative. Senator Hillary Clinton (D-NY) took a climate impacts tour of Alaska in 2005, having earlier called climate change an economic opportunity that America, "the most innovative, creative nation in the world," could certainly cope with (Clinton 2003).

More interesting were several post-Hurricane Katrina bills introduced by Senators Barack Obama (D-IL), Jay Inslee (D-WA), Evan Bayh (D-IN), Richard Lugar (R-IN) and Tom Harkin (D-IL). The Obama-Inslee bill, "Health Care for Hybrids," would boost fuel efficiency amongst US automakers by offering discounts on employee and pension health care costs in exchange for industries' investment of half of the savings into the development and manufacture of fuel efficient vehicles. Bayh's bill, the "Vehicle and Fuel Choices for American Security Act," would require the White House and Federal agencies to develop a plan to reduce US oil consumption by 2.5 million barrels a day by 2015, and by ten million barrels a day by 2031.[16] Half of the ten Senate co-sponsors of the Bayh bill were Republicans, as were 22 of 26 co-sponsors of a similar House bill. This support stemmed in large part from a highly influential effort by an unusual coalition of conservative Republican hawks and environmentalists, known as the Set America Free campaign, launched in 2005. The campaign's goals were to reframe US energy policy towards reducing or eliminating American dependence on imported oil (SAF 2006).

Also in 2006, Senators Richard Lugar (R-IN), Tom Harkin (D-IA), and Barack Obama (D-IL) introduced the "Fuel Security and Consumer Choice Act" that would require all US marketed vehicles to be Flexible Fuel Vehicles (FFVs) within ten years with 10 percent to be FFVs within 18 months of the bill's passage. FFVs can use regular gasoline or ethanol based fuel, the latter reducing GHG emissions from automobiles considerably. In 2005, Senator Lugar had spearheaded the 2005 Fuels Security Act to increase ethanol, biodiesel and cellulosic biomass production in the US, a goal incorporated in modified form into the 2005 energy bill (Griscom Little, 2005b).

Conclusion: US Climate Change Norms and Discourses in the Early 2000s

The foregoing presents an overview of the innovative ways that progressive climate change policy norms and discourses continued to develop in the US in the 2000s despite the position of the G.W. Bush Administration. The analysis illustrates that the pathways for change were numerous and varied, and that diverse actors used new

16 US oil consumption in 2005 was roughly 20 million barrels a day.

and ongoing justifications for action. Most notably, the framings for climate action developed by burgeoning discourse coalitions increasingly stressed the economic opportunities of climate protection action and the economic costs of inaction. Frames about job creation and competitiveness became increasingly common, as did emphasis on the creation and dominance of markets for emission reduction technologies. In essence, the economic opportunity narrative was a "value conversion" reframing (Rochon 1998). An issue previously thought to present only costs was instead reframed as posing significant economic benefits. Rural communities joined this gathering discourse coalition by emphasizing the financial benefits of on-farm or forest-based carbon sequestration operations. Economic need for climate protection-derived income, due to constrained state budgets, represented a related storyline, as did the potential savings (translated into profits or additional budgetary resources) from energy efficiency.

A similarly significant change occurring during this period was the framing of US energy independence as a national security issue. Clearly, rising oil prices and difficulties with the war in Iraq greatly influenced this emerging storyline. But environmental leaders took advantage of the national mood to forge new collaborations with those previously perceived as opponents. The Set America Free coalition was the most significant example of an environmental discourse coalition painting a "value connection" that forged links between phenomena previously understood as relatively unrelated (Rochon 1998). Right wing Republicans were brought into progressive climate change coalitions by a framing of US-based biofuels production as necessary to US national security. Accepting their positioning in a security (rather than a climate change) narrative, this coalition nonetheless fostered a common political project from previously divergent organizations (Hajer 1995).

A final significant climate change storyline in the early 2000s was evident in the US religious communities' framing of climate change as a moral issue and environmental problems as spiritual ones. The "What would Jesus Drive" campaign represented the apex of a much broader shift occurring among progressive churches. Framing GHG emission reductions as a "new way to love your neighbor" infused the US religious community debate on climate issues with a new depth. The frame represented a discursive value connection in the way that it brought moral reflection as taught by Jesus into the narrative (Rochon 1998). Frames emphasizing that GHG emission reductions would also decrease local air pollutants relied on similar moral framings (asthma had reached serious proportions in US inner cities by 2006) as well as an appeal to economic common sense.

Against these frames, opponents of progressive climate change norms primarily mobilized storylines emphasizing the economic costs of climate protection actions. Discourse here emphasized climate policies as amounting to regressive taxes, likely to result in increased energy and health care costs, and problems with international competitive advantage. Efforts to turn these storylines around, through emphasizing the economic opportunities of climate protection, struggled onward for discursive predominance in the early and mid-2000s, reflecting perhaps the difficulty predicted by Rochon in achieving a discursive "value conversion" or an essential reversal of the frame of the economic burden of climate change into a frame of economic opportunity. But the staggering human cost and economic devastation of hurricane Katrina, coupled with the increasing number of high profile corporations taking action on climate change, put a new human face on climate change economic calculations, saying in essence: this is what climate change destruction and climate change protection look like. Where do you stand?

In general, the diffusion of the progressive climate change norms and policies described above seems to represent a mix of what Cass (this volume) calls "constitutive norms" or norms that reflect an actor's true internalization of concern about an issue, and "regulatory norms," norms that are adopted by actors under coercion. Only in the case of corporations were shaming campaigns openly deployed. There is no doubt that public concern about climate change increased in the early 2000s. If, despite this, politicians did cynically develop the legislative proposals at the federal level discussed in this chapter due to a sense of political expediency (that is to say, a need to have a progressive position on climate change in order to survive politically) then we can expect the seriousness and number of politicians taking these stances to fluctuate in coming years if public opinion swings in a new direction. In the case of corporations, economic concerns raised by investors and insurers helped shape GHG emission reduction actions as both constraining and constitutive norms. Demanding new data sets and analyses, insurers and investors changed the incentive structures that corporations faced related to climate change. As corporate leaders learned more, however, they may have also become more concerned about climate change and more aware of the potential economic opportunities associated with GHG emission reduction actions and green branding (framing the discourse much like Paterson and Stripple discuss in their chapter). Movements by corporations away from serious climate-related actions in the years following the period discussed in this chapter would indicate that internalization of norms did not occur, but rather that actions were adopted as a form of temporary green-branding or, what some would call "greenwash."

Not discussed in this chapter, but representing another significant shift in US culture and discourse around climate change in the early 2000s, was the increase, in late 2005 and 2006, of on-line carbon offset retail businesses such as CarbonFund and TerraPass. Selling carbon offsets to consumers and corporations alike, these businesses represented the future of the just emerging corporate goal of "carbon neutrality." That this new norm was touching the international stage indeed reflected the huge strides forward in US domestic climate change discourses since 2001, when G.W. Bush proclaimed the "death" of the Kyoto Protocol.

The most notable development in the early 2000s discussed in this chapter was the rapid and broad-based increase in the number of state, regional and city government entities, corporations and even universities and churches that adopted and acted on relatively ambitious GHG emission reduction targets. Many of these entities discovered cost savings as they did so. Some sub-national US GHG emission reduction targets explicitly referenced international GHG emission reduction norms as contained in the Kyoto Protocol (the US Mayors Initiative), while state norms mirrored less well known, but still international norms, such as those articulated in the United Kingdom by Prime Minister Tony Blair for 60 or 70 percent GHG emission reduction targets over 50 years. Sub-national "local" actors and institutions repositioned themselves as international actors, joining international discourse coalitions calling for and implementing significant GHG emission reductions. By joining together, cities' mayors laid claim to an international stage and debate. What this case suggests is that while the Kyoto Protocol may indeed fade away, the norms and values that it reflects most certainly will not.

References

Anonymous (2006), Investors Turn Up Heat on Insurers to Disclose Risk from Climate Change. *Insurance Journal*. <http://www.insurancejournal.com>.

Anonymous (2005), *Executive Order #S-3-05* and *Governor Establishes Greenhouse Gas Emission Reduction Targets*. California Climate Change Portal. <http://www.climatechange.ca.gov/>.

Anonymous (2002), *US Climate Action Report*. US Government Printing Office. Washington D.C.

American Legislative Exchange Council (2004), *Sons-of-Kyoto' Legislation: States React to the Myth of Global Warming*. <http://www.alec.org/news>.

________. (2006), *Environmental Federalism*. <http://www.alec.org/2/natural-resources/talkingpoints>.

Aston, A. and Helm, B. (2005), The Race Against Climate Change: How top companies are reducing emissions of CO2 and other greenhouse gases. *Business Week*. December 12. <http://businessweek.com/magazine/content/05_50/b3963401.htm>.

Austin Energy (2006), *Creating a Market for PHEV's around the Country: 50-CityPlan*. <http://www.austinenergy.com>.

Barringer, F. and Wald, M. (2005), G.E. Chief Urges US to Adopt Clearer Energy Policy. *New York Times*. May 9, Section C, p. 2.

Bulkeley, H. and Betsill, M.M. (2005), *Cities and Climate Change: Urban Sustainability and Global Environmental Governance*. New York: Routledge.

Bumiller, E. and Nagourney, A. (2006), Bush, Resetting Agenda, Says US Must Cut Reliance on Oil. *New York Times*. February 1. <http://www.nytimes.com>.

California Air Resources Board (2006), *Backgrounder: The Greenhouse Effect and California*. <http://www.arb.ca.gov/cc/factsheets/ccbackground.pdf>.

Carbon Disclosure Project (2005), Carbon Disclosure Project 2005: On behalf of 155 investors with assets of $21 trillion. <http://www.cdp.org>.

CERES (2005), *Availability and Affordability of Insurance Under Climate Change: A Growing Challenge for the US* <http://www.ceres.org/pub>.

Chicago Climate Exchange (2006), *Members of the Chicago Climate Exchange*. <http://www.chicagoclimatex.com>.

Clean Air Cool Planet (2006), Campuses for Climate Action <http://www.cleanair-coolplanet.org/for_campuses.php> [accessed December 11, 2006].

Clinton, H. (2003), *Excerpts from Sen. Hillary Clinton's Floor Speech on First Day of Senate Debate on the McCain-Lieberman Climate Stewardship Act*. <http://www.environmentaldefense.org/article.cfm?contentid=3334>.

Consultative Group on Biological Diversity (2006), *Climate and Energy Funders*. <http://www.cgbd.org>.

Egan, T. (2006), The Greening of America's Campuses. *New York Times*. January 8, Section 4A, p. 20.

Environmental Protection Agency (2006a), *Climate Leaders: Partner's Goals*. <http://www.epa.gov/climateleaders/partners/ghggoals.html>.

Environmental Protection Agency (2006b), Companies Set Aggressive Greenhouse Gas Emission Reduction Goals *EPA Press Release* (October 12) <http://www.epa.gov/climateleaders/news/epareleases.html>.

Evangelical Environmental Network (2006), *Global Warming Briefing for Evangelical Leaders*, and *What Would Jesus Drive? Frequently Asked Questions*. <http://www.creationcare.org>.

Goodstein, L. (2006), Evangelical Leaders Join Global Warming Initiative. *New York Times*. February 8, p. 16.

Griscom Little, A. (2006), Mass Backward: Mass. lawmakers pushing to join climate pact, despite Romney's objections. *Grist On-line Magazine*. January 26.

________. (2005), Shooting the Moon: The Apollo Alliance's grand vision for energy independence is a distant legislative goal, but it can help transform politics right now. *Grist On-line magazine*. October 5. <http://www.grist.org>.

________. (2005a), Amend and Hallelujah: Climate finally getting more notice in Senate with energy bill amendments. *Grist On-line Magazine*. May 26.

________. (2005b). "The School of Barack: Obama and a bipartisan crew of colleagues unveil eco-friendly bills on energy. *Grist On-line Magazine*. November 22.

Hajer, M. (1995), *The Politics of Environmental Discourse: Ecological Modernization and the Policy Process*. Oxford: Clarendon Press.

Hall, K.G. (2006), Administration backs off Bush's vow to reduce Mideast oil imports. Knight Ridder. Washington, D.C. February 1. <http://www.commondreams.org/headlines06/0202-05.htm>.

Ignatius, D. (2005), Corporate Green. *Washington Post*, May 11. <http://www.washingtonpost.com>.

Interfaith Climate Change Network (2006), *Religious Leaders Statement on Energy Conservation*, and *Driven By Values* campaign. <http://www.protectingcreation.org>.

Interfaith Power (2006), *California Interfaith Power and Light: Spiritual Leaders Get Active on Energy*. <http://www.interfaithpower.org>.

International Council of Local Environmental Initiatives (2005), *Clinton Announces ICLEI's Climate Commitment at Inaugural Meeting of Clinton Global Initiative*, and *Cities for Climate Protection Program*. <http://www.iclei.org>.

Kehrl, B. (2005), *States team up on environment*. December <http://www.stateline.org/>.

Marzilli, J. (2005), *Laboratories of Progress: Its time to look past the blockage in Washington and fight for good energy policy at home*. American Prospect. October 5. <www.prospect.org>.

NEG/ECP (2001), *Climate Change Action Plan*. [Online]. <http://www.negc.org/documents/NEG-ECP%20CCAP.PDF>.

Pacific Gas and Electric (2006), *PG&E Proposes Program to Give Customers an Opportunity to Help The Environment by Reducing Greenhouse Gas Emissions: Voluntary Climate Protection Program First-of-its-Kind in the Nation for Utility Customers*. <http://www.pge.com/news/news_releases/q1_2006/060125.html>.

Pew Center. (2006), *Illinois Establishes Agricultural Credit Program*. State and Local News. <http://www.pewclimate.org>.

________. (2005), *Learning from State Action on Climate Change: November 2005 Update*. <http://www.pewclimate.org>.

Pope, C. (2005), *A New Environmentalism*. American Prospect. October 5.. <http://www.prospect.org>.

Rabe, B. (2002), *Greenhouse and Statehouse: The Evolving State Government Role in Climate Change*. Pew Center on Global Climate Change. <http://www.pewclimate.org>.

Rainforest Action Network (2006), *The Global Finance Campaign: Ending Destructive Investment*. <http://www.ran.org>.

Rochon, T. (1998), *Culture Moves: Ideas, Activism and Changing Values*. Princeton: University Press.

Rutenberg, J. (2006), Solution to Greenhouse Gases is New Nuclear Plants, Bush Says, *New York Times*. May 24. <http://www.nytimes.com>.

Seattle gov. (2005), *The US Mayors Climate Protection Agreement*, and *Who is Involved*. <http://www.ci.seattle.wa.us/mayor/climate>.

Set America Free (2006), *Blueprint for Energy Security*. <http://www.setamericafree.org>.

Shellenberger, M. and Nordhaus, T. (2004), *The Death of Environmentalism: Global Warming Politics in a Post-Environmental World*. The Breakthrough Institute. <http://www.thebreakthrough.org/images/Death_of_Environmentalism.pdf>.

The Climate Group (2006), *Carbon Down, Profits Up*. <http://www.theclimategroup.org>.

Tufts Climate Initiative (2005), *EPA Declares Tufts Climate Initiative Tops in Climate Protection*. <http://www.tufts.edu/tie/tci/EPAaward.html>.

Union of Concerned Scientists (2005), *Renewable Energy Expected from State Standards and Funds*. <http://www.ucs.org >.

United States Conference of Catholic Bishops (2004), *Letter to US Senate on Climate Change*. <http://www.nrpe.org/issues/i_air/air_catholic01.htm>.

West Coast Governors' Global Warming Initiative (2003), *Background*. <http://www.climatechange.ca.gov/westcoast/documents>.

________. (2004), *West Coast States Strengthen Joint Climate Protection Strategy*. <http://www.climatechange.ca.gov/westcoast/releases>.

Western Governors' Association (2004). *Western Governors Launch Initiative to Spur Clean, Diversified Energy in the West; Govs. Richardson, Schwarzenegger to Lead Effort*. <http://www.westgov.org>.

PART II
Discourse Analytical Perspective

Chapter 6

Climate Governance Beyond 2012: Competing Discourses of Green Governmentality, Ecological Modernization and Civic Environmentalism

Karin Bäckstrand and Eva Lövbrand

Introduction[1]

It is today commonly argued that international climate politics is moving into a new era. With the entry into force of the Kyoto Protocol in February 2005, many years of multilateral negotiations resulted in a legally binding agreement with quantitative targets that commit industrialized countries to reductions in national greenhouse gas emissions until 2012. However, the route set out by the Kyoto Protocol is far from uncontested and contemporary post-2012 negotiations speak of new ways of making climate policy in the future. In this debate the notion of a stringent multilateral agreement with negotiated emission targets and timetables is increasingly challenged by decentralized, market-oriented and partnership-based narratives calling for more flexible, cost-effective and participatory means to manage anthropogenic climate change.

This chapter adopts a discursive framework in order to critically analyze the policy rhetoric permeating debates on contemporary and future climate governance. Climate governance in this context refers both to the multilateral negotiations and the institutional framework associated with the United Nations Framework Convention on Climate Change (UNFCCC) and the Kyoto Protocol, as well as deregulated governance modes involving sub-state and non-state actors in various partnerships and multi-sectoral networks. The aim is to identify how central discourses have interacted and competed over meaning in the climate negotiations, and how this discursive struggle is manifested in the post-2012 debate on future governance

---

1 Portions of this chapter appear in an earlier version in Bäckstrand, K. and Lövbrand, E. (2006) Planting Trees to Mitigate Climate Change. Contested Discourses of Ecological Modernization, Green Governmentality and Civic Environmentalism, *Global Environmental Politics* 6 (1), pp. 50–75, and are reprinted with permission.

options. We argue that the narratives emerging in the climate domain mirror central trends in global environmental governance: local is pitted against global, North vs. South, public vs. private, decentralization vs. centralization, and economic efficiency vs. environmental integrity. We connect our conception of the three central meta-discourses of global environmental governance (green governmentality, ecological modernization and civic environmentalism) with the micro-discourses surfacing in the climate field.

Green governmentality refers to a science-driven and centralized multilateral negotiation order, associated with top-down climate monitoring and mitigation techniques implemented on global scales. Ecological modernization, on the other hand, represents a decentralized liberal market order that aims to provide flexible and cost-optimal solutions to the climate problem. The civic environmentalism discourse includes radical and more reform-oriented narratives that challenge and resist the dominance of the two former discourses. The radical version of civic environmentalism advocates a fundamental transformation of Northern consumption patterns and abandonment of capitalism and state-centric sovereignty to realize a more eco-centric and just world order. In contrast, the reform-oriented version highlights how the vital force of a transnational civil society, which serves as a complement to state-centric practices, can increase the public accountability and legitimacy of the climate regime.

The first section of the chapter presents key tenets of the discursive framework by conceptualizing the power-knowledge relationships at play in the articulation of global environmental discourses. In the following three sections, we spell out the main characteristics of the three overarching discourses of global environmental governance and indicate how they are reflected in the climate domain. Finally, we examine shifts in these discourses prompted by discussions on future "institutional architectures" for the climate regime and reflect upon the revitalized discursive struggle following these post-2012 talks. We argue that three discourses have proliferated in different phases of the climate negotiations. Whereas the adoption of the UNFCCC in the early 1990s was driven by the command and control logic of green governmentality, the climate debate has, since the signing of the Kyoto Protocol in 1997, become strongly dominated by the market-driven and cost-effective narratives of ecological modernization. Through its promise of mutual benefits for public and private actors in North and South, the international carbon market enacted by the Kyoto Protocol's three flexibility mechanisms has effectively silenced the equity and burden-sharing imperatives embedded in the civic environmentalism discourse. However, we argue that the post-2012 debate has propelled the discursive contestation over future climate governance options and, once more, opened up the process for a struggle over meaning in the climate domain. This discursive struggle includes broader normative debates on North-South fairness, burden-sharing, poverty alleviation, participatory democracy and sustainable development in the making of future climate governance.

Climate Change and Discourses Analysis

Discursive analysis has gained ground and proliferated in the study of global environmental change in the intersecting fields of sociology, political ecology and policy studies (Hajer 1995; Litfin 1994). A central insight of this disparate work is to identify power relationships associated with dominant narratives surrounding "environment" and "sustainable development." In the following section we highlight four dimensions of discourse analysis that are prominent in the literature and relevant for our study. Firstly, discourses are conceived of as a shared meaning of phenomena. Global environmental change in general and climate change in particular are permeated by a struggle over meaning and symbolic representation. In line with Hajer (1995, 45), we understand discourses as "specific ensembles of ideas, concepts and categorization that are produced, reproduced and transformed in a particular set of practices."

Secondly, the exercise of power is closely tied to the production of knowledge, which in turn can sustain a discourse. Hence, discourses are embedded in power relations, "as historically variable ways of specifying knowledge and truth—what is possible to speak at a given moment" (Ramanzanglo 1993, 19). Discourses as "knowledge regimes" bring us squarely to the role of science. In expert-driven global environmental change research, scientific knowledge, techniques, practices and institutions enable the production and maintenance of discourses (these points are further addressed in the chapter by Lahsen). Thirdly, in line with argumentative discourse analysis, we subscribe to a conception of discourse that bridges the gap between the linguistic aspects and institutional dimensions of policy-making. In this vein, discourse analysis can be brought to the forefront of the analysis of power and policy. Policies are not neutral tools but rather a product of discursive struggles. Accordingly, policy discourses favor certain descriptions of reality and hereby empower certain actors while marginalizing others. The concept of discourse institutionalization is useful as it refers to the transformation of discourse into institutional phenomena (Hajer 1995, 61).

Finally, we align ourselves with a discourse analysis that includes the notion of agency. Recent studies have advanced concepts such as "discourse coalitions" and "knowledge broker" to highlight how agents are embedded in discourses (Hajer 1995; Litfin 1994). From this perspective discourses are inconceivable without discoursing subjects or agents that interpret, articulate and reproduce storylines congruent with certain discourses. We employ the concept of discursive agent and argue that political power stems from the ability to articulate and set the terms of a discourse. To conclude, we employ a discursive framework that sheds light on how discourses are deeply embedded in scientific practices and techniques, institutionalized in global policy arenas and articulated by agents spanning the public-private and global-local divide. In the sections below, we present three discourses that arguably underpin both policy practice and academic debates of environmental governance. They also provide rough maps for understanding the discursive framing of the climate problem. Hence, we spell out how these meta-discourses have materialized in climate governance both with regard to the Kyoto framework and future policy scenarios beyond 2012. However, as will be demonstrated, each discourse is heterogeneous and thus is being constantly changed and redefined. Consequently, there are overlaps and conflicts between the discourses when we seek to make sense of environmental governance in general and climate change in particular.

Green Governmentality—Managing the Global Climate

Since the mid 1980s, climate governance has been associated with a science-based monitoring of the global climate system and a multilateral climate regime negotiated under the United Nations' auspices. Through the work of an expanding climate science community during the second half of the twentieth century, anthropogenic climate change at an early stage was defined as a global problem caused by human fossil fuel burning and elevated levels of greenhouse gases in the atmosphere. This global outlook was reflected in the first assessment of the Intergovernmental Panel on Climate Change (IPCC) in 1991, and provided the rationale for the "global politics of the climate" that today is manifested by the UNFCCC and the Kyoto Protocol (Miller 2004, 55; see also Paterson and Stripple's chapter in this volume). Together, these two multilateral accords have established climate change as an intergovernmental issue that is best managed through a worldwide monitoring and reporting system for national sources and sinks of greenhouse gases. The EU has been the most vocal agent of this discourse and is commonly viewed as a global normative leader in promoting environmental multilateralism in general (Lightfoot and Burchell 2005). Accordingly, the EU has consistently advocated a UN-sponsored climate regime based on mandatory and verifiable targets and timetables for domestic emission cuts, where industrialized countries take the lead (Vogler and Bretherton 2006; Grubb et al 1999).

We interpret this centralized multilateral administration of the climate problem as an expression of a green governmentality discourse. Green governmentality epitomizes a global form of power tied to the modern administrative state, mega-science and the business community. It entails the administration of life itself, including individuals, population and the natural environment (Moss 1998, 3). Governmentality is a broad concept introduced by Michel Foucault in order to distinguish the particular mentalities, or rationalities, of government and administration in early modern Europe (Dean 1999). In Foucault's work, governmentality refers to a new form of power exercised by the modern state over its population. According to this framework, the security of the state is not primarily guaranteed by a control over territory, but rather by monitoring, shaping and controlling the people living on that territory (Darier 1999, 23). However, in contrast to traditional conceptions of government that presume that it is possible to regulate human behavior rationally towards certain ends, governmentality involves a multiplicity of authorities and disciplinary mechanisms that together foster and restrict the possibilities of individual and collective identities (Darier 1999, 17; Oels 2005). Through the related concept of bio-politics, Foucault discussed how these "techniques of power" extend to all aspects of human life, including birthrates, health and hygiene, and accordingly, regulate the behavior of entire populations (Foucault in Darier 1999, 22; Dean 1999).

The green interpretation of governmentality extends this optimization of life to the entire planet and the very biosphere in which people live (Dean 1999, 99). In the name of sustainable development and environmental risk management, a new set of administrative truths and knowledges has developed that enables a human stewardship over nature and an all-encompassing management of its resources (Rutherford 1999). Scholars have connected the range of science advisors on the environmental arena with the construction and articulation of these new "eco-knowledges" (Luke 1999; Miller 2004). During the past decades, a growing number of experts have been engaged in the monitoring and management of nature, and human patterns of pollution and environmental degradation. According to Luke (1999), these "eco-managerialist" practices classify and legitimize appropriate ways of dealing with the environment and thus enforce "the right disposition of things" between humans and nature. Recent initiatives such as "sustainability science" (Kates et al 2001) and "earth system science" (Schellnhuber et al 2005; Steffen et al 2004) exemplify the far-reaching ambitions built into this science-based administration of global environmental change. Together, these international research programs approach nature as a coherent life support system or "planetary machinery" (Steffen et al 2004, 9) in need of scientific monitoring and management. The general presumption is that a comprehensive mapping of the earth system will allow science to guide nature-society interactions along sustainable trajectories (Crutzen 2002).

In the field of climate change, this planetary expression of bio-politics is particularly pronounced. Satellite supervision of the Earth's vegetation cover, advanced computer modeling of atmospheric and oceanographic processes, a global grid of meteorological stations and carbon flux towers represent examples of the technical infrastructure used by expert groups to study, monitor and predict human-induced climate change. Influenced by this scientific ambition to map and manage the climate system, the UNFCCC and the Kyoto Protocol today rest upon an extensive monitoring and reporting scheme that commits all ratifying states to regular inventories of greenhouse gas emissions from a range of societal sectors (for example, energy, industry, agriculture, forestry and waste). Since the entry into force of the UNFCCC in 1994, industrialized states have submitted annual greenhouse gas inventories in accordance with a standardized set of methodologies and reporting guidelines developed in cooperation with the IPCC (IPCC 2000, 1997). In parallel to these annual inventories, all parties to the UNFCCC are also obliged to report national policies and measures adopted to implement the convention in "national communications" (UN 1992, Article 4.1, 12). Until now, developed states have submitted three comprehensive reports, with a fourth on its way, whereas most developing countries have completed (or are in the process of completing) their first national communications (UNFCCC 2006). In order to administer the vast range of data and information produced through these extensive monitoring and reporting practices, the institutional framework organized around the UNFCCC secretariat in

Bonn has grown and turned into the global hub around which the multilateral climate order revolves.

In its far-reaching expression, green governmentality has been viewed as a top-down discourse that effectively marginalizes alternative understandings of the natural world (Fogel 2003). Through a detached and powerful view from above, a "global gaze," nature is approached as a terrestrial infrastructure subject to state protection, management and domination (Litfin 1997; see also Heather Smith in this volume). The science-driven and sovereignty-based UNFCCC and Kyoto process is in this discourse framed as an instrumental and technocratic project embedded in expert-oriented and publically inaccessible storylines that favor policy and research elites (Boehmer-Christiansen 2003).

However, a more reflexive vision of green governmentality has been proposed in which dominant narratives of planetary management are replaced by an attitude of humility and self-reflection (Jasanoff 2003; Lash et al 2000; Litfin 1997). In this reflexive version of the discourse, global policy elites acknowledge local complexities and invite local actors to participate in the creation of just and credible institutions (Fogel 2003). The idea of a more reflexive governmentality is to some extent detected in recent work of the IPCC. Apart from the panel's explicit ambition to broaden its expertise beyond Northern science networks, regional scale modeling activities and local vulnerability studies have introduced new physical and social complexities to climate science that today challenge the IPCC's original focus on atmospheric processes and global mean temperatures (Watson 2002). In the preparatory work for the fourth assessment report, due in spring 2007, the IPCC has begun to explore the many links between climate change, local development and adaptation efforts, and has thereby initiated a process to redress the climate problem in a more contextualized sustainable development storyline (IPCC 2005).

A similar discursive shift can be detected in more recent negotiation rounds. Since COP 7 in Marrakesh in 2001, the EU's emphasis on a global climate mitigation regime has become increasingly challenged by vulnerability and adaptation narratives that seek to place the climate problem in the context of local and regional development efforts in the South (Tschakert and Olsson 2005; Ott et al 2004; Jacob 2003; Najam et al 2003). At Marrakesh the G77 members managed to bring about three new funds that aim at strengthening developing countries' capacity to adapt to the impacts of climate change and reducing local vulnerabilities in the South. These include the Special Climate Change Fund, the Least Developed Country Fund and the Adaptation Fund. At COP 10 in Buenos Aires in December 2004, a "Program of Work on Adaptation and Response Measures" was initiated to continue the work on impacts, vulnerability and adaptation to climate change, a decision that was strengthened at COP11/MOP1 in Montreal in 2005 (Müller 2006). Adaptation was further consolidated as an emblematic issue at COP12/MOP2 in Nairobi in 2006, but was not matched by substantial commitment (Pew Center 2006). Accordingly, these attempts to complement the science-driven and multilateral focus on global emission targets with local adaptation efforts in the South has not made any fundamental imprints on the first commitment period of the Kyoto Protocol (2008–2012). As we will discuss in subsequent sections, the adaptation and vulnerability narratives instead tie into a discursive struggle over future climate policy options beyond 2012.

Ecological Modernization—Flexible and Cost-Effective Climate Governance

In parallel to the large-scale multilateral administration of environmental degradation proposed by green governmentality, a discourse of ecological modernization has gained ground in Western industrial societies since the 1980s. The distinct feature of this discourse is the compatibility between economic growth and environmental protection, or more specifically between a liberal market order and sustainable development (Bernstein 2001). The premises of this liberal narrative are that environmental degradation can be decoupled from economic growth, and that capitalism and industrialization can be made more environmentally friendly through green regulation, investment and trade (Hajer 1995). Building upon the notion of mutually reinforcing links between economic growth and environmental protection proposed by the Brundtland Commission in the mid 1980s (WCED 1987), ecological modernization emerges as a win-win discourse in relation to the neo-Malthusian doom and gloom narrative of the 1970s.

According to Christoff (1996), ecological modernization over time has come to represent a continuum of weak and strong versions. The weak version builds upon a technocratic and neo-liberal economic storyline that represents technological adjustment and innovation as key to "eco-efficiency" and cost-effectiveness. Emanating from the experience of advanced industrialized countries, it calls for innovative scientific techniques for integrated pollution control, market-driven strategies to internalize environmental costs, and a changing role of government towards more flexible, decentralized, cost-effective and collaborative policy-making.

The strong ecological modernization discourse has emerged within academic debates on risk society (Beck 1992) and reflexive modernization (Beck et al 1994), and adopts a critical approach to dominant policy paradigms and modern institutions in the management of environmental threats. It moves beyond a Euro-centric perspective on clean production and instead highlights the equity and justice dimensions embedded in the environment-development nexus of sustainable development debates (Carter 2001, 214; Eckersley 2004, 75). Central to strong ecological modernization is the idea of "good governance" (Bernstein 2001) entailing participation, deliberation and inclusion of civil society and stakeholders in environmental policy processes, which resonates with reform narratives in the discourse of civic environmentalism (see next section). For the purpose of this chapter, we equate the discourse of ecological modernization with the "weak expression" since it is arguably the predominant discourse in global policy rhetoric and practice. Dressed in this manner, ecological modernization has been understood as "a process of continually improving environmental productivity by means of new technologies and management practices" (Eckersley 2004, 73).

Flexible and cost-effective environmental problem solving is a central discursive feature in the weak version of ecological modernization that has been widely incorporated into the international climate regime. These narratives are manifested by the global trade in carbon credits organized around the Kyoto Protocol's three flexibility mechanisms: international emissions trading (IET), joint implementation (JI) and the Clean Development Mechanism (CDM). The idea to commodify the climate by creating a market on which carbon credits can be bought and sold emerged as a reaction against the anticipated costs of curbing domestic GHG emissions in industrialized countries. The US can be seen as the main discursive agent for ecological modernization by its advocacy for market and trade-based solutions to address climatic change and, paradoxically, the architect behind the Kyoto mechanisms (Müller 2005, 1). A central storyline promoted by US delegates in the pre-Kyoto negotiations in the mid 1990s was that climate mitigation investments in developing countries will result in greater relative greenhouse gas benefits than the same investment in the North (Matthews and Paterson 2005; Grubb et al 1999). Providing industrial states with the flexibility to purchase emission reductions from countries where these reductions can be carried out at a lower economic cost was thus framed as the most cost-efficient way to come to terms with the climate problem (Matsou 2003; Repetto 2001; Trexler and Kosoff 1998). While initially resisted by developing states, as well as the EU and environmental non-governmental organizations (NGOs), the idea of a global carbon market gained ground in the multilateral process after the Kyoto meeting in 1997. Today the "cap-and-trade" system enacted through the flexibility mechanisms has become a widely accepted part of "the Kyoto concept."

Since January 2005, the ecological modernization agenda is further manifested through the EU Emission Trading Scheme (ETS). The ETS was designed to help the EU member states reach their emission reduction targets under the Kyoto Protocol, and is legitimized by the same cost-efficiency and flexibility storyline as the Kyoto carbon market. The general presumption is that the market forces will guide participants towards the cheapest emission reductions in the EU region and hereby give them the flexibility to choose their own optimal solutions (EEA 2006; CAN 2006). In the first pilot phase running until 2007, the EU trading scheme is limited to CO_2 emissions from the power and heat sectors, oil refineries, the pulp and paper, metal and cement industries in the EU member states. However, in the second phase, starting in 2008, the ETS participants will also be able to buy carbon credits earned through JI and CDM projects (EU 2004). It has also been suggested that the ETS should link its trading activities with actors and institutions in the US (Biermann 2005, 286). Although the Bush administration decided to pull out from the Kyoto process in spring 2001, a range of US private and sub-state actors have explored the opportunities associated with the emerging carbon market. One of the most recent voluntary initiatives at the sub-state level is the "Regional Greenhouse Gas Initiative" (RGGI), a cap-and-trade regime involving seven North-eastern states in the US (see a discussion on this initiative in Fogel's chapter).

At first glance the deregulated and market-based agenda of ecological modernization challenges the state-centric and science-based multilateral negotiation order associated with the green governmentality discourse. The flexibility mechanisms have opened up innovative markets for new branches of business and engaged a range of private actors including carbon traders, carbon finance specialists, carbon auditors and verifiers in climate governance (European Commission 2004). The stepped-up support for Research and Development (R&D) in low-carbon technologies has also turned leading renewable energy companies into central actors around which a new round of de-coupled economic growth will be organized (Matthews and Paterson 2005; Paterson 2001c, 12). The almost 300 CDM projects currently registered by the CDM Executive Board have spurred a range of public-private partnerships spanning the North-South divide (URC 2006). According to Oels (2005), this plethora of voluntary commitments and public-private partnerships marks a distinct shift from the centralized command and control logic of green governmentality. Ecological modernization, or advanced liberal government, today provides the dominant discursive space in which climate change is interpreted and conceptualized (ibid.). A similar argument is made by Bernstein (2001, 118), who suggests that the Kyoto Protocol's carbon market epitomizes the normative shift in environmental governance towards a liberal economic order.

However, recent negotiation rounds have suggested that deregulated and partnership-based modes of networked climate governance are compatible with and even dependent upon the political decisions, rules and policies negotiated in the multilateral climate regime. While the private sector actors present at COP11/MOP1 in Montreal in 2005 criticized the complex and costly registration process developed for CDM projects, they also put pressure on the Kyoto parties to create a more long-term regulatory framework that will secure corporate investments in the carbon market beyond 2012 (Wittneben et al 2006; Müller 2006). These calls for clear terms of trade point to the artificial character of the Kyoto market and the close links between the green governmentality and ecological modernization discourses. Instead of challenging the institutional framework provided by the UNFCCC and the Kyoto Protocol, the global carbon market's flexible and cost-efficient solutions to the climate problem are highly dependent upon negotiated caps on national emissions and regulatory science for carbon monitoring, verification and accounting. As noted by Peeters (2003, 154), the effectiveness of any emissions trading scheme will largely depend upon its compliance and enforcement provisions. Although a discursive fault line exists between the science-driven and state-centric governance practices built into green governmentality, and the deregulated flexibility ideal of ecological modernization, we suggest that these two discourses have become mutually constitutive in climate governance.

Civic Environmentalism—From Radical Equity to Pragmatic Reform

While the entry into force of the Kyoto Protocol in February 2005 was celebrated by the multilateral community and most environmental NGOs, the neo-liberal framing of the climate regime associated with the market-based flexible mechanisms has been subject to much critique and controversy, not least on moral and ethical grounds. During the past decade the dominant policy discourses of green governmentality and ecological modernization have been challenged by counter-narratives aimed at redefining the basic principles of climate governance towards equity and ecological sustainability. We propose that these counter-narratives are interwoven in a civic environmentalism discourse that is reflected in wider debates of environmental governance. This discourse includes radical and more reform-oriented narratives that differ in their diverging views on the role of the sovereign state, the multilteral system and the capitalist economy in addressing the global climate problem.

Radical Resistance The radical edge of civic environmentalism is deeply skeptical of contemporary environmental governance practices. Departing from a neo-Gramscian perspective, this discourse emphasizes the relations of power and powerlessness as the core of international institutions and negotiation processes. It is informed by a radical ecology agenda that advocates a fundamental transformation of consumption patterns and existing institutions to realize a more eco-centric and equitable world order. In addition, it contests the structures of global environmental governance that revolve around the liberalization of markets, free trade and sovereignty based practices (Elliot 2002). For example, the multilateral financial institutions and UN agencies are criticized for being underpinned by a neo-liberal bias in their quest for market-oriented policies, privatization and deregulation at the expense of environmental protection. Likewise, the enduring power structures of sovereignty, capitalism, scientism, patriarchy and even modernity generate and perpetuate the environmental crisis while consolidating structural inequalities between the global North and South (Parks and Roberts 2006; Paterson 2001a). Hence, the radical resistance discourse suggests that social movements should challenge and resist inequitable power structures which currently define the global institutional framework (Dryzek 2006; Mason 2005; Lipschutz 2004). Examples to be resisted include partnership agreements and voluntary commitments which represent the retreat of the state, the rise of transnational corporate power and capital accumulation, and mask the relationships of power and domination embedded in global environmental politics (Paterson 2006, 66; Elliot 2002).

The radical resistance version of civic environmentalism challenges the neo-liberal approaches to climate governance found in the discourse of ecological modernization. From the radical perspective, the Kyoto negotiations have followed a "neoliberal script" (Byrne et al 2004, 443) that "parcels up the atmosphere and establishes a routinized buying and selling of 'permits to pollute'" that distracts the attention away from rich countries' unequal over-consumption of the world's resources (Bachram 2004, 2). By establishing a market of "world carbon dumps," the Kyoto mechanisms tie into colonial patterns that serve the economic interests of Northern states and corporations at the expense of the poor and vulnerable in the South (Lohman 2005, 204). The radical resistance discourse also embodies a critique of the expert dominance embedded in the green governmentality discourse.

According to the fiercest critics of the IPCC and the Kyoto Protocol, the large uncertainties associated with greenhouse gas accounting amount to a "scientific fraud" and signifies the rise of "carbon technocracy" (ibid., 30).

Climate equity and climate sustainability are two narratives that predicate this discourse. The climate equity narrative concerns the fair distribution of the costs associated with the mitigation of climate change among states, as well as the compensation to poor countries for their disproportionate vulnerability to severe climate events (Ott et al 2004). This narrative was advocated by developing countries in the negotiations preceding the United Nations Conference on Environment and Development (UNCED) in the early 1990s, and resulted in the incorporation of equity concerns into the UNFCCC. According to the first principle of the convention, countries should protect the climate system "on the basis of equity and in accordance with their common but differential responsibilities and respective capabilities" (UN 1992, Article 3.1). The convention hereby recognizes that industrialized states should take the lead in combating climate change due to their historical responsibility for the build-up of greenhouse gases, while considering developing countries' special needs and vulnerabilities.

The equity imperative was epitomized in the heated debate between the US-based think-tank World Resources Institute (WRI) and India's Center for Science and Environment (CSE) in the early 1990s, in which "luxury emissions" generated by the materialist lifestyle in rich industrialized countries were pitted against "survival emissions" resulting from subsistence living in the South (Mwandosya 2000; Agarwal and Narain 1993). Closely related to this argument are the recurrent proposals for per capita assignments as the only equitable approach in the allocation of emission reductions among states (Pinguelli-Rosa and Munasinghe 2002; Agarwal and Narain 1991; see also the discussion in the Paterson and Stripple chapter). Per capita allocations are based on the ethical principle that all people have equal rights to the atmospheric commons. As a consequence, each state should be assigned a greenhouse gas target that corresponds to its share of the world's population. According to the principle of distributive justice, large emitters in the North will thus have to make considerable cuts in domestic greenhouse gas emissions before developing countries with low per capita emissions will be asked to do so (Blok et al 2005; Ott et al 2004). While per capita assignments surfaced in the pre-Kyoto negotiations as the most convincing ethical ground for the distribution of responsibility in the climate domain (Paterson 2001b, 124), the proposal was downplayed in policy rhetoric and practice after the signing of the Kyoto Protocol in 1997 and in the allocation of quantitative targets to industrialized countries. However, as will be discussed in the subsequent section, the per capita model has resurfaced in the post-2012 debate as a critique of the Northern calls for "broadened participation" by developing countries in a future mitigation scheme (Biermann 2005; Blok et al 2005; Ott et al 2004).

The climate sustainability narrative criticizes what is conceived as the largest shortcoming of the Kyoto Protocol: how to meet the long-term goal to stabilize atmospheric greenhouse gases at "safe levels." According to the radical critique, the climate regime primarily is focused on the expansion of the global carbon economy by means of short-term and arbitrary cuts in national emission far below the 60–80 percent cuts in CO_2-equivalents proposed by the IPCC (Najam et al 2003, 25). Since the Kyoto Protocol will not solve the climate change problem, the resistance discourse entails a return to the original "sustainability" mandate of the UNFCCC. In a recent publication of the Climate Action Network (CAN 2006), the critique extends to the EU emissions trading scheme. The CAN notes that the emission caps set by the member states during the first pilot phase of the ETS have been a great disappointment. Although most of the EU15 countries are lagging behind their Kyoto commitments, the ETS allows many European industries to increase rather than reduce their emissions (CAN 2006, Michaelowa, Butzenegeiger and Jung 2005). Climate Action Network (CAN) suggests that this concession to corporate interests in the allocation of emission caps diminishes the EU's international credibility as a climate change leader (CAN Europe 2006).

The Reformist Discourse In contrast to its radical cousin, the reformist version of the civil environmentalism discourse, also branded "participatory multilateralism" (Elliot 2004), stresses how the vital force of a transnational civil society and global deliberative politics can complement state-centric practices (Betsill 2006; Dryzek 2006). The UNCED process in the early 1990s invited participation by so-called "major groups" in environmental negotiations through various stake-holding practices and consultative fora. While the sovereignty norm and the legal notion of states as exclusive decision-making authorities in international negotiations remain, the active participation of non-state actors (such as business, NGOs and women's groups) has created more pluralistic, complex, "glocal" and multifaceted governance arrangements (Friedman et al 2005; Fisher and Green 2004; this trend is also apparent in the Fogel chapter). The civil society forum at the Rio Conference emerged as a template for subsequent summits on environment, development and poverty. Likewise, the partnership agreements (entailing public-private voluntary agreements) and "multi-stakeholdership" at the World Summit on Sustainable Development held in Johannesburg in 2002 have served increasingly as a model for collaborative problem solving in issues such as sustainable development, energy, water and climate in the twenty-first century (Bäckstrand 2006; Streck 2004; Andanova and Levy 2003).

The reform-oriented version of civic environmentalism suggests that increased access and stakeholder participation of international diplomacy outside the community of legitimate decision-makers (i.e., states) can increase the public accountability and legitimacy of multilateral institutions. Increased civil society participation brings specialized expertise to international negotiations and links global agendas with local concerns (Lipschutz 1996; Wapner 1996). The reform agenda subscribes to a pluralistic global environmental order and affirms the rise of public-private partnerships between NGOs, business and governments as they hold the promise of result-based environmental problem solving (Hale and Mauzerall 2004; Benner et al 2003). While there is disbelief that the market alone can generate an equitable distribution of resources or halt environmental degradation, cross-sectoral cooperation between market, state and civil society actors is encouraged. Increased civil society participation is predicted to raise the green profile of the global economic order, trade agendas and multilateral financial institutions. Moreover, it is suggested that the plethora of partnerships between business and NGOs will speed up the greening of the business agenda (Joyner 2005; Streck 2004).

Whereas the radical critique of the neo-liberal underpinnings of the climate regime long dominated the civic environmentalism discourse in the climate domain, a turning point came at the resumed COP6 in Bonn in July 2001 after the US withdrawal from the Kyoto negotiations. The previously strong opposition against the Kyoto mechanisms turned into acceptance by large segments of the NGO community as well as vulnerable states in the Alliance of Small Island States (AOSIS) coalition. The prospect of the Kyoto Protocol being "dead" prompted pragmatic efforts to save the agreement with the EU in particular exercising leadership in the negotiations (Vogler and Bretherton 2006). Even though the trenchant critique of the marketization and commodification of climate politics remains within the NGO community, a reform-oriented version of civic environmentalism has emerged that admits the potentials of the global carbon market and the multi-sectoral voluntary partnerships enabled by the Kyoto mechanisms and the EU emission trading scheme (Gulbrandsen and Andresen 2004). In line with the reflexive green governmentality and strong ecological modernization narratives, this discourse focuses on locating climate change within the wider sustainable development agenda, North-South equity and intergenerational justice (Najam et al 2003, 228; Wiegant 2001). Human development plays a central role in the reform agenda (Pan 2005) as well as institutional innovations that can meet both the social development goals in the South and the Northern call for cost-effective climate mitigation. The reform discourse highlights potential synergies between the Climate Convention, the Convention of Biodiversity, and the Convention of Desertification as well as Agenda 21 and the Johannesburg Plan of Implementation (Tschakert and Olsson 2005). In addition, the Climate Change Fund for capacity building and the Less Developed Country (LDC) fund to aid the poorest countries in climate change adaptation are conceived as important steps towards greater North-South equity (Ott et al 2004).

The reformist version of civic environmentalism diverges from the radical discourse by conditionally embracing the so-called "co-benefits" embedded in the Kyoto market (Blok et al 2005). According to this line of argument, CDM and JI projects can serve as a bridge between industrialized and developing countries as well as with economies in transition (Streck 2004). These projects hold the potential to generate sustainable development in the host country by channeling new investments and cleaner energy technologies, promoting sustainable forest management practices, stimulating local employment and building institutional capacity (UNDP 2006). By bridging the global-local divide, they may also enhance public-private stakeholder participation and thus increase the accountability and legitimacy of the climate regime (Tschakert and Olsson 2005; Holm Olsen 2005). However, in contrast to the win-win rhetoric in the ecological modernization discourse, the reformist discourse underlines that sustainable development synergies only can be realized if norms of equity, participation and accountability inform the global carbon market (Elliot 2004, 118; Gupta 2002). Therefore, the reformist discourse calls for increased transparency and civic participation in the implementation of the Kyoto mechanisms (CAN 2006; Ott et al 2004).

Discursive Struggle in the Post-2012 Debate

During the past decade and a half, a discursive struggle has played out in the climate change domain, and different ideas on how to understand and manage anthropogenic climate change have proliferated and competed over salience. All three discourses outlined above have played a significant part in the formation of the current climate regime. However, as indicated in the sections above, the narratives underpinning international climate politics are in constant redefinition and change. At the onset of the climate negotiations in the early 1990s, planetary control of atmospheric greenhouse gas concentrations was in focus. Prompted by the global climate scenarios in the IPCC's first assessment report, states adopted a multilateral accord, building upon the top-down and administrative rationality of green governmentality. The environmental stewardship ambitions embedded in this discourse provided both the rationale for continued investments in climate science, and paved the way for the global GHG monitoring, reporting and verification apparatus that currently functions as an essential part of the UNFCCC and Kyoto Protocol implementation review.

In parallel with this initial manifestation of green governmentality, the radical discourse on North-South equity and "luxury emissions" in industrialized countries also made a significant imprint on the emerging climate regime. Its emphasis on differentiated responsibilities between North and South shaped the negotiations during the 1990s, and provided the rationale for the allocation of quantitative emission reduction targets among industrialized countries in Kyoto in 1997. However, through the adoption of the Kyoto Protocol, the radical equity agenda in the civic environmentalism discourse has gradually been marginalized in favor of liberal narratives of ecological modernization. During the past decade, political focus has revolved around the Kyoto mechanisms and the trade in carbon credits enabled by the emerging carbon market. Resting upon a salient win-win rhetoric, these market mechanisms have functioned as a discursive compromise through their promise to simultaneously deliver climate protection, technological innovation, economic growth and sustainable development.

Along with recent discussions on a second commitment period for the Kyoto Protocol beyond 2012, we see a new wave of discursive contestation over future climate governance options. This turning point in climate politics has broadened the scope of climate governance by including treaty and non-treaty based agreements, multilateral tracks as well as private-public arrangements (Michaelowa et al 2005). A more pluralistic, or fragmented, climate governance order is now in the making. Possible elements in future climate governance include the Kyoto framework, bilateral agreements, domestic voluntary programs, public-private partnerships, technology-based agreements, and carbon trading within and beyond the Kyoto market (Blok et al 2005; Bodansky 2004; Aldy et al 2003). The emerging climate order can be captured in terms such as "public-private multilateralism" or "market multilateralism," where corporations and non-profit organizations increasingly shape and transform multilateral practices (Bull and McNeill 2006). Three entry points are prominent for the renewed discursive struggle in the post-2012 era: 1) mandatory target setting and burden-sharing, 2) the efficacy of the UN-based institutional framework for climate change mitigation, and 3) the long-term environmental integrity of the climate regime.

Who Should Pay the Costs?

The distribution of costs and burdens associated with climate change is a divisive issue in the post-2012 talks. As previously discussed, the political focus on the Kyoto mechanisms has consolidated the predominance of the ecological modernization discourse and its market-based, flexible and cost-effective greenhouse gas mitigation approach. Radical calls for distributive justice and per-capita assignments were effectively silenced by the great win-win compromise embedded in the global carbon market. However, in the post-2012 debate, radical narratives have once more surfaced in relation to the Northern calls for "broadened participation" by developing countries in a future mitigation scheme (Müller 2005). The proposition that large emitters in the developing world (such as China, India and Brazil) should accept mandatory emission reduction targets has revitalized questions of North-South equity and the industrialized countries' historical responsibility for the build-up of atmospheric GHG concentrations. Recurrent proposals for per capita assignments according to the "contraction and convergence" model (Blok et al 2005; Michaelowa et al 2005; Bodansky 2004; see also the Paterson and Stripple chapter) reflect the demands for establishing a more equitable North-South bargain in the post-2012 climate regime.

However, the vast differences among developed and developing countries with respect to emission trends, income levels and vulnerability to climate impacts have turned climate equity into a more complex ethical issue in the post-2012 debate (Bruggink 2003, 279). Proposals for more systematic differentiation on the basis of historic responsibility for elevated atmospheric GHG concentrations, financial capability to pay for mitigation measures, intra-national equity among rich and poor social groups, and differentiation between luxury and survival emissions have therefore surfaced in the debate (Biermann 2005, 282; Blok et al 2005; Ott et al 2004).

"Multi-stage" approaches have also been introduced in order to gradually involve developing countries in a future emission target scheme (Den Elzen et al 2006; Blok et al 2005). The EU is a key discursive agent in forging a North-South climate pact based on gradual engagement of developing countries in a future cap and trade system, combined with agreements on technology transfer and financial support for adaptation measures and more sustainable land use practices (Blok et al 2005). In these post-2012 negotiations, the climate equity storyline has come to involve a stronger focus on adaptation strategies for vulnerable groups and areas, which was reaffirmed at the recent COP12/MOP2 in Nairobi , also coined the "Africa adaptation COP" (IISD 2006, 18). Although adaptation has always been an integral part of the climate convention (UN 1992, Article 3), advocates of the civic environmentalism discourse have aspired to turn it into a core concern rather than a second order issue in relation to future mitigation commitments (Yamin 2005). Increased funding for adaptation measures in the South, beyond the relatively limited resources provided by the Kyoto Protocol's adaptation funds, is one of the central issues in this debate (Ott et al 2004).

Although the equity imperatives in the civic environmentalism discourse are gaining ground in the post-2012 talks, calls for climate equity continue to be contested. The US non-ratification and ensuing critique of the "fundamentally flawed" design of Kyoto Protocol partly stems from a resistance against the discursive rationale of differentiated responsibilities between industrialized and developing countries. The recurrent argument, most clearly articulated by President George W. Bush in spring 2001, is that "major population centers" such as India and China will contribute the larger share of global GHG emissions within the coming decades. Freeing developing countries from mandatory emission targets will jeopardize industrialized countries' efforts to stabilize climate change at safe levels, give growing economies in the South unfair competitive advantage, and hence bring serious harm to the US economy (Schreurs 2004; Lisowski 2002). As one of the most prominent articulators of the ecological modernization discourse, the Bush administration has opposed the assignment of multilateral targets beyond 2012 and has instead continued to stress the power of markets, technological innovation and public-private partnerships (Abraham 2004). The rejection of the Kyoto process by the Bush administration reflects a more general skepticism towards multilateralism, which is a considerable source of transatlantic tension between the US and the EU (Vogler and Bretherton 2006; Scheuers 2004).

Centralization, Decentralization or Both?

The US promotion of financial incentives and voluntary agreements between governments, private actors and NGOs is a second entry point for discursive contestation in the post-2012 talks. In line with the flexible and decentralized logic of ecological modernization, the US position has emerged as a strong counter-narrative to the centralized rationale of the green governmentality discourse. Whereas the EU members' insistence on an inclusive multilateral regime, that commits the UN member states to long-term targets and timetables for emission reductions, has been framed as costly, ineffective and inherently Eurocentric (Kellow 2006). In contrast, US negotiators have portrayed domestic initiatives as well as bilateral and public-private partnerships for technology innovation as key to sustained economic growth and future energy and climate security (Watson 2002).

The decentralized, pluralistic and technology-led modes of governance proposed by this ecological modernization discourse are today widely reflected in the many "institutional architectures" suggested for the post-2012 regime (Blok et al 2005; Aldy et al 2003). Although the multilateral framework embedded in the green governmentality discourse still seems to dominate academic and policy debates on future governance options (Bodansky 2004), cost-effectiveness, market-based instruments and technology partnerships have surfaced as central discursive traits in the ongoing post-2012 talks under the UN auspice (Egenhofer 2005; IISD 2005). The discourse is strongly influenced by the various technological partnerships initiated by the US government during recent years, including "the International Partnership for a Hydrogen Economy" (April 2003), "the Carbon Sequestration Leadership Forum" (June 2003), "the Methane to Markets Partnership" (July 2004), and "the Asia-Pacific Partnership on Clean Development and Climate" (AP6) between Australia, China, India, Japan, South Korea and the United States (July 2005) (see Philibert 2005). Together these initiatives epitomize the US focus on voluntary commitments and "diversified approaches" in the post-2012 debate. They also signify the rise of networked governance in global environmental politics (Schreurs 2005, 359 ff).

However, despite the apparent conflict between the top-down, UN-based and mandatory mitigation rationale embedded in the green governmentality discourse and the bottom-up, decentralized and voluntary climate initiatives proposed by ecological modernization, a future compromise is in the making where multilateral treaty making complements public-private partnerships and bilateral treaties, domestic action and voluntary target setting (Michaelowa 2005). At COP11/MOP1 in Montreal in 2005, a "dual track" was initiated involving both a continued cap-and-trade system in accordance with the Kyoto Protocol, and a dialogue on long-term cooperative action with the Kyoto skeptics in the US administration (Wittneben et al 2006). Outside the direct negotiation context, business actors have increasingly come to approach the Kyoto trade in carbon credits as a new site of capital accumulation in which firms can invest and make profits (Matthews and Paterson 2005; see also Paterson and Stripple this volume). The corporate interest in the global carbon market and the new investments in carbon-efficient technologies challenge the argument that mitigation commitments necessarily harm economic growth and competitiveness (Azar 2005). Instead they point to a potential discursive compromise between the EU members' multilateral agenda and the Bush administration's voluntary and deregulated approach in future climate governance.

Long-Term Climate Stabilization or Short-Term Economic Gains?

This brings us to the third and final arena of discursive struggles in the post-2012 era. The contemporary dominance of the ecological modernization discourse in climate politics is increasingly criticized for weakening the environmental effectiveness of the climate regime. Even if the Kyoto parties would manage the unlikely and meet the average 5 percent reduction goal in 2012, critics claim that the environmental performance of the Kyoto Protocol is modest. The sharpest discursive contestation is in regards to the efficacy of the market-based climate regime (currently epitomized by the EU carbon emission trading scheme) in delivering the cuts in GHG emissions necessary to prevent dangerous human interference with the climate system.

The post-2012 debate has prompted a critical review of the promise of ecological modernization and its short-term focus on cost-minimization and flexibility for industrialized countries. Challenging the predominant neo-liberal framings, key narratives in the civic environmentalism discourse are returning to the policy arena. These include demands for long-term emission reduction commitments beyond 2012 in order to restore the environmental integrity of the climate regime (Najam et al 2003; Pan 2005). The currently debated 2 degree centigrade temperature target, introduced by the EU and several environmental NGOs, suggests that global GHG emissions may need to be reduced by as much as 75 percent over the next 50 years (Azar 2005, 312; see the Pettenger chapter for the Netherlands' support for this position). Such a far reaching commitment may challenge the win-win promise of ecological modernization and, once again, bring trade-offs between short-term economic growth and long-term climate mitigation to the fore.

However, between the eco-modernist promotion of carbon markets and economic cost-effectiveness, and the radical demand for a fundamental restructuring of the global political economy, we argue that the reformist version of civic environmentalism has emerged as a new discursive consensus. In recent negotiation rounds the EU members have aimed for a compromise deal that balances environmental effectiveness, fairness and implementability of the climate regime by gradually phasing in large developing states in a mandatory emission scheme, while at the same time providing synergies between technological innovation, market based strategies and adaptation efforts in the South (Blok et al 2005).

This tendency to reframe climate governance in a manner that has "something in it for everyone" (Blok et al 2005, 9) reflects a growing predominance of reform narratives in the climate policy domain. The representation of climate mitigation *and* adaptation as parallel processes in recent negotiation rounds and in the IPCC's fourth assessment report suggests that this moderate version of the civic environmentalism discourse is emerging as a new bargain between North and South (Najam et al 2003, 228). Capacity building, increased investments in social, economic and technical resilience for the most vulnerable countries and new multi-sectoral public-private partnerships are, in this discourse, framed as effective means to simultaneously ensure climate security, social development, poverty eradication, participatory democracy and enhanced North-South dialogue. Accordingly, we suggest that this socially grounded win-win narrative, which converges with both the reflexive green governmentality and strong ecological modernization discourses, represents the new discursive compromise in post-2012 climate politics. Currently, the EU represents the main discursive agent of this emerging global climate pact embedded in sustainable development and adaptation narratives. There have been proposals that the EU should continue to exercise diplomatic leadership and build a "climate coalition of willing," that non-ratifying states such as the US and Australia will eventually be tempted to join (Christoff 2006).

However, the extent to which the reform-oriented version of the civic environmentalism discourse will be able to reorient climate governance practices is yet to be discovered. While contemporary climate policy undoubtedly has a strong (neo)liberal framing, with carbon capitalism, trade and investment as key elements, the making of a future climate regime along the lines of the reform agenda will involve a balancing act between multiple goals of environmental integrity, market efficiency, human development and North-South equity. Whether such a balancing act will fulfill its promise to deliver a more coherent, sustainable and legitimate climate regime will be of central interest for environmental scholars in the coming years.

References

Abraham, S. (2004), The Bush Administration's Approach to Climate Change. *Science*, 305, pp. 616–617.

Agarwal, A. and Narain, S. (1991), Global Warming in and Unequal World: A Case of Environmental Colonialism. *Earth Island Journal*, (Spring), pp. 39–40.

Aldy, J., B. Scott and Stavis, R.N. (2003), Thirteen Plus One: A Comparison of Global Climate Policy Architectures. *Climate Policy*, 3, pp. 373–397.

Andonova, L. and Levy, M. (2003), "Franchising Global Governance: Making Sense of the Johannesburg Type II Partnerships" in O.S. Stocke and Ø.B. Thomessen, eds. *Yearbook of International Co-operation on Environment and Development 2003/2004*. London: Earthscan, pp. 19–31.

Azar, C. (2005), Post-Kyoto Climate Policy Targets: Costs and Competitiveness Implications, *Climate Policy*, 5, pp. 309–328.

Bachram, H. (2004), Climate Fraud and Carbon Colonialism: The New Trade in Greenhouse Gases. *Capitalism Nature Socialism,* 15 (4), pp1–16.

Beck, U. (1992), *Risk Society: Towards a New Modernity.* London: SAGE Publications.

Beck, U., A. Giddens and Lash, S., eds. (1994), *Reflexive Modernization: Politics, Traditions and Aesthetics in the Modern Social Order*. Stanford: Stanford University Press.

Benner, T., C. Streck and Witte, J.M., eds. (2003), *Progress or Peril? Networks and Partnerships in Global Environmental Governance: The Post-Johannesburg Agenda*. Berlin and Washington D.C.: Global Public Policy Institute.

Bernstein, S. (2001), *The Compromise of Liberal Environmentalism*. New York: Colombia University Press.

Betsill, M. (2006), "Transnational actors in international environmental politics" in M. Betsill, K. Hochstettler and D. Stevis, eds. *International Environmental Politics*. Basingstoke: Palgrave MacMillan.

Biermann, F. (2005), Between the USA and the South: Strategic Choices for European Climate Policy. *Climate Policy*, 5, pp. 273–290.

Blok, K., N. Höhne, A. Torvanger and Janzic, R. (2005), *Towards a Post-2012 Climate Change Regime: Final Report*. European Commission, DG Environment, Brussels.

Bodansky, D. (2004), *International Climate Efforts Beyond 2012: A Survey of Approaches*. Pew Center on Global Climate Change.

Boehmer-Christiansen, S. (2003), Science, Equity and the War against Carbon. *Science, Technology and Human Values,* 28, pp. 69–92.

Bruggink, J.J.C. (2003), "The Multi-Sector Convergence Approach to Global Burden Sharing of Greenhouse Gas Reductions" in M. Faure, J. Gupta and A. Nentjes, eds. *Climate Change and the Kyoto Protocol: The Role of Institutions and Instruments to Control Global Change.* Cheltenhamn & Northhampton: Edward Elgar, pp. 279–291.

Bull, B. and McNeill, D. (2006), *Development Issues in Global Governance: Public-Private Partnerships and Market Multilateralism*. London and New York: Routledge.

Byrne, J., L. Glover, V. Inniss, I. Kulkarni, Y. Mun, N. Toly and Y. Yang (2004), "Reclaiming the Atmospheric Commons: Beyond Kyoto" in V. Grover, ed. *Climate Change: Five Years After Kyoto*. Enfield, NH: Science Publishers, pp. 429–452.

Bäckstrand, K. (2006), Multi-stakeholder Partnership for Sustainable Development: Rethinking Legitimacy, Effectiveness and Accountability. *European Environment*, 16 (5), pp. 290–306.

Carter, N. (2001), *The Politics of the Environment: Ideas, Activism, Policy*. Cambridge: Cambridge University Press.

Christoff, P. (2006), Post-Kyoto? Post-Bush? Towards an effective climate coalition of willing. *International Affairs,* 82 (5), pp. 831–60.

________. (1996), Ecological Modernization, Ecological Modernities. *Environmental Politics*, 5 (3), pp. 476–500.

Climate Action Network (CAN) (2006), *National Allocation Plans 2005–7: Do they Deliver? Key Lessons for Phase II of the EU ETS.* <http://www.climnet.org/EUenergy/ET/NAPsReport_Summary0306.pdf>.

________. Europe (2006), Use the ETS Tool – Reduce Now. *Hotspot*, 41.

Crutzen, P.J. (2002), Geology of Mankind. *Nature*, 415, p.23.

Darier, É. (1999), "Foucault and the Environment: An Introduction" in É. Darier, ed. *Discourses of the Environment.* Oxford: Blackwell Publishers, pp. 1–34.

Dean, M. (1999), *Governmentality: Power and Rule in Modern Society.* London, Thousand Oaks, New Delhi: Sage Publications Ltd.

Den Elzen, M., M. Berc, P. Lucas, and XX (2006), Multi-stage: A Rule-based Evolution of Future Commitments Under the Climate Change Convention. *International Environmental Agreements*, 6 (1), pp. 1–28.

Dryzek, J.S. (2006), *Deliberative Global Politics*. Cambridge, UK: Polity Press.

Eckersley, R. (2004), *Green State: Rethinking Democracy and Sovereignty.* Cambridge: MA, MIT Press.

Egenhofer, C. (2005), Technology in a post-2012 Transatlantic Perspective. *CEPS Policy Brief*, 86, (November 2005).

Elliot, L. (2002), "Global Environmental Governance" in R. Wilkinson and S. Hughes, eds. *Global Governance: Critical Perspectives*. London & New York: Routledge, pp. 57–74.

________. (2004), *The Global Politics of the Environment*. London: Macmillan Press Ltd.

European Commission (2004), *EU Emission Trading: An Open Scheme Promoting Global Innovation to Combat Climate Change*. Information brochure. Belgium: European Communities.

European Environment Agency (EEA) (2006), *Using the Market for Cost-Effective Environmental Policy*. EEA Report No 1/2006, Copenhagen, Denmark.

European Union (EU) (2004), *Directive 2004/101/EC of the European Parliament and of the Council of 27 October 2004. Amending Directive 2003/87/EC Establishing a Scheme for Greenhouse Gas Emission Allowance Trading within the Community, in Respect of the Kyoto Protocol's Project Mechanisms*. <http://ec.europa.eu/environment/climat/emission/pdf/dir_2004_101_en.pdf>.

Falkner, R. (2005), American Hegemony and the Global Environment. *International Studies Review*, 7 (4), pp. 585–599.

Fisher, D. and Green, J. (2004), Understanding Disenfranchisement: Civil Society and Developing Countries Influence and Participation in Global Governance for Sustainable Development. *Global Environmental Politics*, 4 (3), pp. 65–84.

Fogel, C. (2003), "The Local, the Global and the Kyoto Protocol" in S. Jasonoff and M.L. Martello, eds. *Earthly Politics: Local and Global in Environmental Governance*. Cambridge, London: The MIT Press, pp. 103–126.

Friedman, E., K. Hochstetler and Clark, A.M. (2005), *Sovereignty, Democracy and Global Civil Society*. New York: State University of New York Press.

Grubb, M., C. Vrolijk and Brack, D. (1999), *The Kyoto Protocol: A Guide to Assessment*. London and Washington: The Royal Institute of International Affairs.

Gulbrandsen, L. and Andresen, S. (2004), NGO Influence in the Implementation of the Kyoto Protocol: Compliance, Flexibility, Mechanisms, and Sinks. *Global Environmental Politics*, 4 (4), pp. 54–73.

Gupta, J. (2002), "Capacity Building and Sustainability Knowledge in the Climate Change Regime" in F. Biermann, S. Camp and J. Jacob, eds. *Proceedings of the 2002 Berlin Conference on the Human Dimensions of Environmental Change "Knowledge for the Sustainability Transition: The Challenge for Social Science"*. Global Governance Project: Amsterdam, Berlin, Potsdam and Oldenburg, pp. 154–164.

Hajer, M. (1995), *The Politics of Environmental Discourse: Ecological Modernization and the Policy Process*. London: Oxford University Press.

Hale, T. and Mauzerall, D. (2004), Thinking Globally, Acting Locally: Can the Johannesburg Partnerships Coordinate Action on Sustainable Development? *Journal of Environmental and Development*, 13 (3), pp. 220–39.

Holm Olsen, K. (2005), *The Clean Development Mechanism's Contribution to Sustainable Development – A Review of the Literature: UNEP Risoe Centre.*

Climate and Sustainable Development. Risoe National Laboratory. Roskilde, Denmark.

International Institute for Sustainable Development (IISD) (2005), Summary of the UNFCCC Seminar on Governmental Experts. *ENB*, 12. p.261.

________. (2006), Summary of the twelfth conference of the parties to the UN framework convention on climate change and second meeting of the parties to the Kyoto protocol: 6–17 November 2006. *ENB*, 12 (13), pp. 1–21. <www.iisd.ca/climate/cop12>.

IPCC. (2005), *Report on the Joint IPCC WGII & WGIII Expert Meeting on the Integration of Adaptation, Mitigation and Sustainable Development into the 4th IPCC Assessment Report*. February 16–18, 2005, St Denis, France.

________. (2000), *Good Practice Guidance and Uncertainty Management in National Greenhouse Gas Inventories*. IPCC/OECD/IEA/IGES, Hayama, Japan.

________. (1997), *Revised 1996 IPCC Guidelines for National Greenhouse Gas Inventories*. IPCC/OECD/IEA, Paris, France.

Jacob, T. (2003), Meeting Review: Reflections on Delhi. *Climate Policy*, 3, pp103–106.

Jasanoff, S. (2003), Technologies of Humility: Citizen Participation in Governing Science. *Minerva*, 41, pp. 223–244.

Joyner, C.C. (2005), Rethinking International Environmental Regimes: What Role of Partnership Coalitions? *Journal of International Law and International Relations*, 1 (1–2), pp89–120.

Kates, R.W., W.C. Clark, R. Corell, J.M. Hall, C.C. Jaeger, I. Lowe, J.J. McCarthy, H.J. Schellnhuber, B. Bolin, N.M. Dickson, S. Faucheux, G.C. Gallopin, A. Grübler, B. Huntley, J. Jäger, N.S. Jodha, R.E. Kasperson, A. Mabogunje, P. Matson, H. Mooney, B. Moore III, T. O'Riordan and U. Svedin (2001), Sustainability Science. *Science*, 292, pp. 641–642.

Kellow, A. (2006), A New Process for Negotiating Multilateral Agreements? The Asia-Pacific Climate Partnership Beyond Kyoto. *Australian Journal of International Affairs*, 60 (2), pp. 287–303.

Lash, S., B. Szerszynski and Wynne, B., eds. (2000), *Risk, Environment and Modernity. Towards a New Ecology*. London: Sage Publications Ltd.

Lightfoot, S. and Burchell, J. (2005), The European Union and the World Summit for Sustainable Development: Normative Power or Europe in Action. *Journal of Common Market Studies* 43 (1), pp. 75–95.

Lipschutz, R.D. (2004), *Global Environmental Politics: Power, Perspectives, and Practice*. Washington D.C.: CQ Press.

________. (1996), *Global Civil Society and Global Environmental Governance: The Politics of Nature from Place to Planet*. New York: State University of New York Press.

Lisowski, M. (2002), The Emperor's New Clothes: Redressing the Kyoto Protocol. *Climate Policy*, 2, pp. 161–177.

Litfin, K. (1997), The Gendered Eye in the Sky: A Feminist Perspective on Earth Observation Satellites. *Frontiers*, 18 (2), pp. 26–47.

________. (1994), *Ozone Discourses: Science and Politics in Global Environmental Cooperation*. New York, Colombia University Press.

Lohman, L. (2005), Marketing and Making Carbon Dumps: Commodification, Calculation and Counterfactuals in Climate Change Mitigation. *Science as Culture*, 14 (3), pp. 203–235.

Luke, T. (1999), "Eco-Manegerialism: Environmental Studies as a Power/Knowledge Formation" in F. Fisher and M.A. Hajer, eds. *Living with Nature. Environmental Politics as Cultural Discourse*. Oxford, New York: Oxford University Press, pp. 103–120.

Mason, M. (2005), *The New Accountability: Environmental Responsibility Across Borders*. London: Earthscan Institute.

Matsuo, N. (2003), CDM in the Kyoto Negotiations: How CDM has Worked as Bridge Between Developed and Developing Countries. *Mitigation and Adaptation Strategies for Global Change*, 8, pp. 191–200.

Matthews, K. and Paterson, M. (2005), Boom or Bust? The Economic Engine Behind the Drive for Climate Change Policy. *Global Change, Peace & Security*, 17, pp. 1–20.

Michaelowa, A., S. Butzengeiger and Jung, M. (2005), Graduation and Deepening: An Ambitious Post-2012 Climate Policy Scenario. *International Environmental Agreements*, 5, pp. 25–46.

Michaelowa, A., K. Tanger and Hasselknippe, H. (2005), Issues and Options for the post-2012 Climate Architecture – An Overview. *International Environmental Agreements*, 5, pp. 5–24.

Michaelowa, A. and Butzengeiger, S. (2005), EU Emission trading: navigating between Scylla and Charybdis. *Climate Policy*, 5 (1), pp. 1–9.

Miller, C. (2004), "Climate Science and the Making of a Global Political Order" in S. Jasanoff, ed. *States of Knowledge. The Co-Production of Science and Societal Order.* London & New York: Routledge, pp. 46–66.

Moss, J. (1998), "Introduction: the Later Foucault" in J. Moss, ed. *The Later Foucault*. London/Thousand Oaks/New Dehli: Sage Publications, pp. 1–17.

Mwandosya, M.J. (2000), *Survival Emissions: A Perspective From the South on Global Climate Change Negotiations*. Dar es Salaam: The Centre for Energy, Environment, Science and Technology.

Müller, B. (2006), *Montreal 2005. What Happened, and What it Means*. Oxford Institute for Energy Studies, EV 35, February 2006.

________. (2005), Climate Change post 2012: Transatlantic Consensus and Disagreement. *Journal of Energy Literature*, 1 (1), pp. 1–10.

Najam, A., S. Huq and Sokona, Y. (2003), Climate Negotiations Beyond Kyoto: Developing Countries Concerns and Interests. *Climate Policy*, 3, pp. 221–231.

Oels, A. (2005), Rendering Climate Change Governable: From Biopower to Advanced Liberal Government. *Journal of Environmental Policy & Planning*, 7, pp. 185–207.

Ott, H.E., H. Winkler, B. Brouns, S. Kartha, M.J. Mace, S. Huq, Y. Kameyama, A. Sari, J. Pan, Y. Sokona, P.M. Bhandari, A. Kassenberg, E. La Rovere and A.A.

Rahman (2004), *South-North Dialogue on Equity in the Greenhouse: A Proposal for an Adequate and Equitable Global Climate Agreement.* Report published by Deutsche Gesellschaft für Technische Zusammenarbeit (GTZ), Germany.

Pan, J. (2005), Meeting Human Development Goals with Low Emissions: An Alternative to Emission Caps for Post-Kyoto from a Developing Country's Perspective. *International Environmental Agreements*, 5, pp. 89–104.

Parks, B.C., Roberts, T.J. (2006), "Environmental and Ecological Justice" in M. Betsill, K. Hochstettler and D. Stevis, eds. *International Environmental Politics*. Basingstoke: Palgrave MacMillan, pp. 329–360.

Paterson, M. (2006), "Theoretical Perspectives on International Environmental Politics" in M. Betsill, K. Hochstettler and D. Stevis, eds. *International Environmental Politics*. Basingstoke: Palgrave MacMillan, pp. 54–81.

________. (2001a), *Understanding Global Environmental Politics: Domination, Accumulation and Resistance*. London: Macmillan.

________. (2001b), "Principles of Justice in the Context of Climate Change" in U. Luterbacher and D.R. Sprinz, eds. *International Relations and Global Environmental Change*. Cambridge, MA: MIT Press, pp. 119–126.

________. (2001c), Climate Policy as Accumulation Strategy: The Failure of COP6 and Emerging Trends in Climate Politics. *Global Environmental Politics*, 1 (2), pp. 10–17.

Peeters, M. (2003), "Legal Feasibility of Emissions Trading: Learning Points from Emissions Trading for Ozone-Depleting Substances" in M. Faure, J. Gupta and A. Nentjes, eds. *Climate Change and the Kyoto Protocol. The Role of Institutions and Instruments to Control Global Change.* Cheltenham & Northampton: Edward Elgar, pp. 147–170.

Pew Center (2006), *Twelfth conference of the parties to the UN framework convention on climate change and second meeting of the parties to the Kyoto protocol, 6–17 November 2006*. Washington DC: Pew Center on Global Climate Change.

Philibert, C. (2005), The Role of Technological Development and Policies in a Post-Kyoto Climate Regime. *Climate Policy*, 5, pp. 291–308.

Pinguelli-Rosa, L. and Munasinghe, M. (2002), *Ethics, Equity and International Negotiations on Climate Change*. Cheltenham, UK: Edward Elger.

Ramanzanglo, C. (1993), "Introduction" in C. Ramzanoglo, ed. *Up Against Foucault. Explorations of Some Tensions between Foucault and Feminism*. London and New York: Routledge, pp. 1–28.

Repetto, R. (2001), The Clean Development Mechanism: Institutional Breakthrough or Institutional Nightmare? *Policy Sciences*, 34, pp. 303–327.

Rutherford, P. (1999), "The Entry of Life into History" in É. Darier, ed. *Discourses of the Environment.* Oxford: Blackwell Publishers, pp. 37–62.

Schellnhuber, H.J., P. Crutzen, W. Clark and Hunt, J. (2005), Earth System Analysis for Sustainability. *Environment*, 47 (8), pp. 11–25.

Schreuers, M. (2004), "The Climate Change Divide: The European Union, The United States and the Future of the Kyoto Protocol" in N. Vig and M. Faure, eds.

Green Giants? Environmental Policies of the United States and the European Union. Cambridge, MA: MIT Press, pp. 207–230.

________. (2005), Global Environmental Threats and a Divided Northern Community. *International Environmental Agreements*, 5, pp. 359–346.

Steffen,W., A. Sanderson, P.D. Tyson, J. Jäger, P.A. Matson, B. Moore III, F. Oldfield, K. Richardson, H.J. Schellnhuber, B.L. Turner II and R.J. Wasson (2004), *Global Change and the Earth System: A Planet under Pressure.* Berlin, Heidelberg, New York: Springer-Verlag.

Streck, C. (2004), New Partnerships in Global Environmental Policy: the Clean Development Mechanism. *Journal of Environment and Development*, 13 (3), pp. 295–322.

Trexler, M.C. and Kosloff, L.H. (1998), The 1997 Kyoto Protocol: What Does it Mean for Project-Based Climate Change Mitigation? *Mitigation and Adaptation Strategies for Global Change,* 3, pp. 1–58.

Tschakert, P. and Olsson, L. (2005), Post-2012 Climate Action in the Broad Framework of Sustainable Development Policies: the Role of the EU. *Climate Policy*, 5, pp. 329–348.

UNFCCC (2006), < www.unfccc.int>.

United Nations (1992), *Framework Convention on Climate Change.* UNEP/IUC/99/9. Bonn. UNFCCC 2006 "National Reports." <http://unfccc.int/national_reports/items/1408.php>.URC (2006) "CDM/JI Pipeline Overview" <http://cd4cdm.org/Publications/GuidanceCDMpipeline.pdf>.

United Nations Development Progamme (UNDP) (2006), *The Clean Development Mechanism. An Assessment of Progress*. UNDP, November 2006.

URC (2006), CDM/JI Pipeline Overview, <http://cd4cdm.org/Publications/GuidanceCDMpipeline.pdf>.

Vogler, J. and Bretherton, C. (2006), The European Union as a Protagonist to the United States on Climate Change. *International Studies Perspectives*, 7, pp. 1–22.

Wapner, P. (1996), *Environmental Activism and World Civic Politics*. Albany: State University of New York Press.

Watson, R. (2002), 'The Future of the Intergovernmental Panel on Climate Change', *Climate Policy,* 2, pp. 269–271.

Wiegandt, E. (2001), "Climate Change, Equity and International Negotiations in Change" in U. Luterbacher and D.R. Sprinz, eds. *International Relations and Global Environmental Change*. Cambridge, MA: MIT Press.

Wittneben, B. W. Sterk, H.E. Ott and Brouns, B.(2006), The Montreal Climate Summit: Starting the Kyoto Business and Preparing for Post-2012. *Journal for European Environmental Planning and Law*, 2, pp. 90–100.

World Commission on Environment and Development (WCED) (1987), *Our Common Future.* Oxford: Oxford University Press.

Yamin, F. (2005), The European Union and Future Climate Policy: Is Mainstreaming Adaptation a Distraction or Part of the Solution? *Climate Policy*, 5, pp. 349–361.

Chapter 7

Singing Climate Change into Existence: On the Territorialization of Climate Policymaking

Matthew Paterson and Johannes Stripple

Introduction

Climate change is often figured in academic and popular discourse alike as a "global" challenge, a threat "beyond borders." This "global" character, often taken as a defining feature of climate politics, makes climate change the paradigmatic global environmental problem. Its presumed globality links the climate issue to a broader discussion within International Relations (IR) about the contemporary role of territory and political boundaries. The flows of people, pollution and money across boundaries are read as indicators of a post-Westphalian world, stipulated and stereotyped as a deterritorialized and borderless political space. Climate change is thus contrasted in this discourse with a spatiality of global politics which is constructed as territorial, the parceling up of the world into discrete political units.

We have known at least since Immanuel Kant that categories provide the basis for knowledge. Steve Smith made this claim explicit in his presidential address at the annual meeting of the International Studies Association in Portland, 2003, referring to an aboriginal tradition of "singing the world into existence."[1] Smith wants to draw attention to the extent that the world we take for granted is dependent on the categories we use to make sense of the world. Our chapter will contend that a dominant element in the social construction of climate change has been a discourse concerning territory and territorialization. The spatiality of climate politics is, in other words, sung into existence. This discourse has been one which has simultaneously legitimized climate politics itself, by rendering it intelligible to state elites to whom discourses of state territorial sovereignty are "master discourses," while, simultaneously, it has shaped and circumscribed the range of possible responses to climate change. The argument is less of the more straightforward one that state elites and others interpret climate threats through nationalist lenses, but rather to explore the ways that such a territorial framing is productive of concrete policies pursued.

1 A revised version of the speech is published in ISQ (Smith 2004).

As a consequence, our chapter does not consider so much the question of agency in contemporary constructions of climate change,[2] i.e. we are relatively silent about "who the singer of climate politics might be," but instead focuses more on the musical architecture (the notes, the score, the instruments) that make a particular song possible. Before outlining the structure of this chapter, let us briefly comment on territory as a social construction and the implications of this for global climate politics. Despite the constructivist turn in IR theory, it has taken some time for the concept of territory to be rethought along constructivist lines (Ruggie 1993, 174). Knight noted more than 20 years ago: "in a sense, territory is not; it becomes, for territory itself is passive, and it is human beliefs and actions that give territory meaning" (Knight 1982, 517).

The silence in IR concerning the status of territoriality has, thankfully, been met by an opposing trend in Geography revolving around a general assertion on the social production of space (Ó Tuathail 1996; Lefebvre 1991; Soja 1989). The designation of the climate, or environment, more generally as a "global issue" (see also H. Smith this volume) has particular histories and politics (Chatterjee and Finger 1994; Ross 1991). Hence, the writing of global environmental space cannot be read in an "objective" manner from the scientific knowledge on the subject but rather is the product of this knowledge combined with particular ideological or discursive projects. Many different aspects come into play here, such as the ambitions of scientists themselves, the post-Chernobyl sense of "global" crisis and the interests of globalizing public and private elites (Dobson 2005; Shiva 1993; Sachs 1993; Merchant 1992). Since the territorial configurations of political space are made to appear as natural, certain political and economic solutions to the climate issue becomes legitimized. That territoriality is a human artifact does not limit its influence on politics; "All territorial organization and its associated process of bordering are social and political constructions and, as such, continue to function as powerful mechanisms of control and ordering—at a variety of spatial and political levels" (Newman 2004). But it does mean that we cannot make simple claims which naturalize the power or identity of particular states by reference to their "territorial size," for example.

---

2 The agency-structure debate has been on going in International Relations (c.f. Dessler 1989; Wendt 1987) with different solutions (c.f. Doty 1997; Bieler and Morton 2001; Suganami 1999) and will not be rehearsed in this context. We conceive of discourses as practices that systematically (re-)create and (re-)produce the object that they speak of. Discourse is not something actors can "deploy'" in a purely strategic sense, it pre-exists them, and structures what they do and how that is interpreted by others (this does not mean people do not try to actively reinterpret the discourses available to them, but they cannot control the effects of those reinterpretations as they are understood by their interlocutors). Colin Hay has framed the question of (and potential for) agency in a useful way: "strategic action is the dialectical interplay of intentional and knowledgeable, yet structurally-embedded actors and the preconstituted (structured) contexts they inhabit" (Hay 1995:2000).

In the following pages, we analyze three specific elements in climate policy discourses that serve to structure concrete policy developments, and illustrate how they reterritorialize climate politics. The first section concerns divergent articulations of the climate issue as one of "danger" or one of "opportunity" that both seem to share an underlying ontology of spatial separation. Secondly, we return to the debates concerning a fair distribution of greenhouse gas (GHG) reductions. Despite claims which base proposed reductions on *individual* entitlements, justice turns out to be about membership in a territorially based political community. Thirdly, we look into the rearticulation of the global carbon cycle as carbon "sinks," a process whereby the flows of carbon are shaped to coincide with the borders of the sovereign state. While it is likely that reterritorializing discourses will continue to shape climate policymaking, there are some interesting contradictions and adjacent political opportunities that our conclusion will explore. In light of these contradictions, climate change can be understood as still an "unsettled" political space and a range of contemporary discourses are involved in the negotiation of a settlement.[3]

Danger and Opportunity

The first element concerns constructions of climate change in terms of the twin poles of danger and opportunity visible in both the climate issue as such as well as in policy measures to deal with it. These contradictory discourses very directly affect the viability or otherwise of measures to limit GHG emissions. So for example, (to simplify matters somewhat) climate policy developments in Europe are shaped by notions of climate change as threat while GHG emissions reductions are constructed as an opportunity—for energy efficiency, balance of payments benefits, technological innovation, and so on. By contrast in the US (hegemonically, at least), climate change is regarded as, at worst, neutral, while climate policy is constructed as a threat to competitiveness, jobs, and sovereignty.

Arrhenius, in his 1896 classic paper "on the influence of carbonic acid in the air ...," did not express concern about the global warming he predicted. With a few exceptions such as G.D. Callendar in the 1930s, such questions did not emerge until the 1970s. Two publications in the early 1970s expressed various ways of conceiving of climate change. The *Report of the Study of Mans impact on Climate* (SMIC) from 1971 emphasized the dangers of rising global average temperatures, while a Central Intelligence Agency (CIA) report from 1974 interpreted available evidence as supporting a trend towards decreasing global average temperatures and envisioned a shortage of food for the poor and the powerless. However, despite the two reports' contradictory assessments (warming vs. cooling) *change* itself is perceived as the source of threat (Wiman and Chong 2000, 19).

---

3 A comparison is found in Fogel's chapter (this volume) on the development of climate change policy norms in the US where she projects a more heterogeneous and contentious picture than what is usually admitted or accounted for.

In the international context, one of the earliest articulations of climate change as a potential "threat" was the first World Climate Conference at Toronto, Canada, in 1979. At the conference, organized by the World Meteorological Association (WMO), scientists called on the world's governments "to foresee and prevent potential man-made changes in climate that might be adverse to the well-being of humanity" (WMO 1979). Some individuals had done this before, such as Callendar in the 1930s and Tickell in 1977 with his book on *Climate Change and World Affairs* (Tickell 1977).[4] The threat orientation continued in much official scientific discourse throughout the Toronto conference where global warming was designated second only to nuclear war in terms of the threat it posed to human society. An excerpt from the 1988 Toronto Statement, speaking of changes in the atmosphere, is characteristic of the age: "…these changes represent a major threat to international security and are already having harmful consequences over many parts of the globe" (Victor and Lanchbery 1995, 32). It was articulated also during that period through the way that climate change was represented in the first series of popular books on the subject (Lyman 1990; Oppenheimer and Boyle 1990; Gribbin 1990; Boyle and Ardill 1989). Perhaps needless to say, NGO statements quickly articulated climate change as a security question and have continued to make that connection. One might also note that public attention to the climate issue came in the US in the aftermath of a series of heat waves and droughts in the summer of 1988. Weart (2003, 155) argues, with regard to these record temperatures and with notice of James Hansen of NASA's (National Aeronautics and Space Administration) timely testimony at a Congressional hearing in late June, that global warming became a question of individual security: "The story was no longer an atmospheric phenomenon: it was a present danger to everyone." Finally, the objective of the climate change framework convention picks up a similar line of thinking with its "to prevent dangerous anthropogenic interference with the climate system" (United Nations 1992, Article 2).

Within the negotiations under the UNFCCC (United Nations Framework Convention on Climate Change), the Alliance of Small Island States (AOSIS) has been the actor that most clearly has articulated climate change as a severe threat to national security and sovereignty. The AOSIS was formed at the second World Climate Conference in 1990 and is an ad-hoc coalition of low-lying states and island states that are particularly vulnerable to sea level rise and extreme weather events. The vulnerability stems from obvious geographical factors and a lack of resources for adaptation measures (Barnett and Adger 2003; Salinger 2000). The articulation of climate change as a threat has also been initiated by the insurance industry. The industry prefers to use the language of risk as it denotes a danger that is insurable, i.e., it can be assessed with some confidence in terms of its probability and potential consequences. A trend of increasing costs of weather-related disasters, and the fear that climate change will aggravate the trend, has brought the insurance industry into the realm of climate change discussions. They have initiated activities in relation to

4 It might in this context to be interesting to note that Tickell was a relative of Aldous Huxley who wrote the influential *Brave New World* in 1932.

the UNFCCC, issued statements and implemented certain financial instruments that have a clear political significance (Jagers and Stripple 2003; Paterson 2001c).

On the broader climate scene, the Pentagon report *An Abrupt Climate Change Scenario and Its Implications for United States National Security* (Schwartz and Randall 2003) has further fuelled a threat oriented discourse, and they situate it squarely and safely within the traditional framework of national security: "We have created a climate change scenario that although not the most likely, is plausible, and would challenge United States national security in ways that should be considered immediately" (Schwartz and Randall 2003, 1). Similar ideas have inspired Hollywood to turn future climate impacts into spectacular dramas with the movies *Waterworld* and *The Day After Tomorrow* and the scenario has also been taken up in the BBC Horizon documentary *The Big Chill* that was first broadcasted in November 2003. While these imaginations of very rapid climate changes are expressed within popular culture (and critiqued within popular culture, as in Michael Crichton's novel *State of Fear*), the established scientific community have also shown increasing interest in the possibility of abrupt climate change. The hypothesis that the climate system can change rapidly, even abruptly, when certain thresholds are crossed was proposed early by Lorenz (1963), Broecker (1987) and Dansgaard (1969), but is now becoming a major research area. One of the best known recent additions to this literature is the US National Research Councils' *Committee on Abrupt Climate Change* (NRC 2002; see also Alley et al 2003) which states that abrupt change is a common feature of the global climate system and has reached up to 10°C in a decade in some regions due to the warming since the last ice age. Another recent articulation of climate change as a threat was the large conference *Avoiding Dangerous Climate Change* organized by the Met Office, UK, for the Department for Environment, Food and Rural Affairs (Met Office 2005). Many of those themes were also debated at a summer meeting at the Aspen Global Change Institute (2005).

Within a focus on the consequences of climate change itself, there is a powerful counter-narrative of "opportunity." This framing was prevalent in the USSR negotiations early on. It is also a recurrent image among the "skeptic scientists" with the point being that when they do not contest whether global warming is occurring, they suggest it might be a good thing. This position holds a respectable lineage—back to Arrhenius who (sitting in Stockholm) thought a long-term climatic warming would be a good thing. Further, a range of "opportunities and risks for human development" is recognized by the IPCC in its assessments of the implications for regions and sectors. Most of arguments revolve around CO_2 fertilization, longer growing seasons, opening up of new agricultural and habitable lands.

At the same time, there is an alternative reading of the "danger/opportunity" constructions, which focuses less on climate change itself and more on the interpretation of proposals to limit emissions that have emerged. Throughout the history of the climate issue a predominant focus of most corporate leaders has been to depict climate policy as a threat to competitiveness, jobs and growth. This position has been most clearly expressed through the Global Climate Coalition, and then articulated most forcefully in the Byrd-Hagel resolution, and further by the

Bush Administration. The position has also been backed by Australia under John Howard (and more recently Canada under Stephen Harper), and it is a discourse that has been consistently articulated by most developing countries (with the exception of Argentina after COP4 in Buenos Aires in 1998). In the International Negotiating Committee (INC) during negotiations that led to the framework convention in 1992, the debate around the principle of the "right to development" illustrates these lines of thought.

By contrast, an articulation of global warming as a business opportunity has been gaining ground. Some corporations (such as those involved in renewable energy, but also oil companies like BP, and some elements of the finance sector) have been active in this regard. The notion of "ecological modernization" has been the key route towards a rearticulation of the relationship between growth and the environment—not as a trade-off, but as potentially complementary (see also Bäckstrand and Lövbrand, this volume). Based politically on the corporatist traditions in continental Europe (particularly Germany, Sweden, the Netherlands and Denmark) which provide a basis for "ecologically modernizing" state intervention—legitimized in neoliberal terms (important in the UK) through arguments by key business guru Michael Porter (see Porter and van der Linde 1995) that environmental regulation can stimulate innovation and improve competitiveness in a global(izing) economy—climate policy becomes interpreted as an opportunity for the "competition state" (Barry and Paterson 2004; Paterson 2001a). In the climate negotiations, the Kyoto Protocol created various mechanisms (Clean Development Mechanism—CDM, Joint Implementation—JI, Emissions Trading) that are designed to create a series of business opportunities through the pursuit of climate policy. The creation of an international carbon market enables companies to profit from the commercialization of low-carbon technologies, products and services, and has been central to the construction of a political coalition favoring Kyoto.

The interplay of these discourses of danger and opportunity has clearly been an important element in shaping the concrete measures and practices that have been developed to deal with climate change. But what interests us here is that, whichever construction becomes hegemonic, the spatiality is territorial. It is national states that are endangered by climate change and, at the same time, national economies which are threatened by or might benefit from action on climate change. Climate change policy in this context is fundamentally constructed through the twin lenses of national security and national economic strategy in a globalizing economy. In both, the state is the "master discourse" which serves to legitimize other discourses.

This reterritorializing discourse is at its most palpable regarding AOSIS as it reiterates traditional security issues (territorial integrity and national identity), but also in the Pentagon report that clearly articulates global warming as a threat to "US national territorial security." Most recently, it has been made highly visible in Canadian (and to a lesser extent Danish, concerning Greenland) discourses of sovereignty in the Arctic, and particularly over the Northwest passage put into question by the receding of Arctic sea ice.

At the same time, reterritorialization of the "danger/opportunity" discourse helps explain the withdrawal of the US from the Kyoto Protocol. Two months after taking office, in March 2001, Bush announced that the United States would withdraw from the Kyoto Protocol. His argumentation echoed the traditional security discourse: "we will not do anything that harms our economy, because first things first are the people who live in America." Later on in June 2001, Bush claimed that "for America, complying with these mandates (the Kyoto Protocol) would have a negative economic impact, with layoffs of workers and price increases for consumers" (Bush, 2001; see Fogel and Cass, this volume, for details on US policy). But conversely, the logic of promoting climate policy for its beneficial economic consequences depends on a nationalist impulse, to garner benefits within a specific territory in a globally competitive economy. This logic is best expressed by the title (and content) of a recent book outlining the "win-win" possibilities in environmental politics generally (and climate change politics specifically)—*The Natural Advantage of Nations* (Hargroves and Smith 2005).

This section has shown how the process of territorialization structures articulations of climate change both as a "danger" and as an "opportunity". The effect of this process has complex political consequences. At times, it legitimates resistance to climate policy, oriented around the threats to the economy and skepticism to interpreting climate change as "threat". At other times, it has enabled climate policy—in particular in some northern European countries—and has provided the basis for promoting such policy for its national economic benefits. Whether it encourages or discourages climate policy, however, the scope of potential legitimations of climate policy is narrowed—a broader discourse of collective responsibility in the face of commonly shared (and produced) threats is eschewed in favor of an account of responsibility for facing danger or taking opportunities which is territorially limited. The next section will illustrate how specific debates concerning justice in relation to climate change are situated within a geography of separated self-contained political spaces, that is to say, territories.

Justice and territory

This limiting of the question of "to whom are we responsible" is further illustrated by examining how the territorial discourse constructs the climate issue in relation to a range of debates around international/transnational justice in relation to climate change. It is no surprise that sovereign territoriality has made its mark on justice within the context of the UNFCCC.[5] For example, the whole "burden sharing" framing of what justice entails is structured in terms of the rights and responsibilities

5 Cass (this volume) outlines in detail differences between *states* concerning their domestic salience of the norm regarding quantified emission reduction targets. Particularly interesting (in the context of our discussion) is his argument that the American government has sought to recast the normative debate to emphasize economic efficiency and global emission reductions rather than national emission reduction commitments.

of states. But it is also there in marginalized discourses and discourses which apparently aim to transcend Westphalian notions of political space. We will, in this section, recapitulate and reconsider how justice and fairness have been articulated in relation to climate change. We will not primarily engage in a normative argument, but instead read the debate on norms for distributing emission reductions through a spatial lens. It is striking how many positions fall into the "territorial trap" (Agnew 1994), either in their very foundations or once action is to be conceived.

Many scholars, negotiators and activists have noted that the human influence on the global climate raises a range of ethical considerations. For example, how much change in climate-related parameters should be tolerated? Is the lack of scientific certainty a legitimate reason for inaction? Does cost-benefit analysis provide a fair grounding for decisions on climate policy? Do developed countries have special responsibilities to act before the poorer nations (cf. Brown 2001)? The climate negotiations have, in particular, brought to the fore debates on distributive justice, for example concerning the allocation of the costs of preventing climate change and coping with the consequences of climate change, as well as what a fair bargaining process and a just allocation of emissions of GHG would be (Paterson 2001b, 120). There are many attempts to cover these vast normative territories (Anand 2003; Rosa and Munasinghe 2003; Toth 1999). Large parts of the contributions to these debates argue for the need and relevance of a specific ethical consideration or the particular merits of a certain approach (Ikeme 2003; Tonn 2003; Rowlands 1997).

The potential answers to the above raised questions can vary considerably. However, let us start to map out how the multilateral negotiations on climate change under the UNFCCC framework have approached the issues. Since climate change is caused by increasing concentrations of atmospheric GHG, the negotiations have from the outset been directed at circumscribing emissions of those gases, mainly carbon dioxide. The question of allocating the distribution of emissions limitations has in practice been organized along one of two paths, both justified publicly in terms of notions of fairness, justice or "equity."[6] Along the first path, state negotiators have started from existing levels of emissions and then haggled over whether to limit them equally or to differentiate emissions limitations according to some rational criteria or simply through crude bargaining. Along the second path, negotiators have argued that emissions should be allocated according to a basic principle of individual equality.

The first of these paths has clearly been most frequently traveled throughout climate politics. It is unambiguously a statist project, which starts from, and reinforces, the conception that it is states that have both rights and responsibilities in relation to the rest of the world. It is fairly obvious that reterritorialization begins here with

---

6 This is despite the variety of ethical and philosophical positions which might underpin different allocation schemes (Paterson 2001b, 1996; Banuri et al 1996; Grubb 1995; Shue 1993, 1992). "Equity" is put in speech marks because of its particularly slippery nature as a term; a whole thesis could be written deconstructing the maneuverings involved in referring to equity rather than the more normatively robust (if essentially contested) term justice.

the fact that it is state negotiators who are mandated to articulate and authorize these accounts of justice. On a somewhat deeper philosophical level, this path entails also the reproduction of the standard ideology of the "sovereign equality of states" that is codified in Article 2 of the United Nations Charter.[7] At the same time, differentiation has taken place within the UNFCCC to accommodate the fact that states are neither equal in their capabilities, nor equally responsible for the build-up of carbon dioxide in the atmosphere. There is, for example, a differentiation between Annex 1 and Annex 2 countries that reflects the types of obligations countries have in relation to the convention and to developing countries.[8] However, there is also a differentiation in emissions reductions targets among Annex A countries in the Kyoto Protocol.

Apart from this official discourse on how to share the burdens of climate change mitigation in an equitable way, there is a marginalized discourse worth exploring both because it is arguably still important to the evolution of the climate regime, and because it articulates a universal potential to counter the simple territorial construction. The initial push in this direction was in the well known paper published by the Centre for Science and Environment in New Delhi in 1990, *Global Warming: a Case of Environmental Colonialism* (Agarwal and Narain 1998). The paper focused on a critique of the World Resources Institute's statistics that had calculated the Third World share of the contributions to the accumulation of atmospheric CO_2 to be 48 percent. Agarwal and Narain's calculations showed that developing countries were only responsible for 16 percent. The rationale of their approach was:

> No country can be blamed for the gases accumulating in the Earth's atmosphere until each country's share in the Earth's cleansing ability has been apportioned on a fair and equitable basis. Since most of the cleansing is done by the oceans and troposphere, the Earth has to be treated as a common heritage of mankind (Agarwal and Narain 1998, 158).

Agarwal and Narain proportionally distributed the world's total sink capacity to each country in relation to its share of the world's population. Thus, "India with 16% of the world's population, gets 16% of the Earth's natural air and ocean 'sinks' for carbon dioxide and methane absorption" (Agarwal and Narain 1998, 158–159, citation in the original). Agarwal and Narain's intervention was an early articulation of a per capita based approach to allocate emission reductions; however their text also included an influential argument about the need to differentiate between subsistence and luxury emissions, i.e. not all emissions are equal and cannot be added together. Overall, when recalculating on a per capita basis, the national contributions of developing

7 Article 2, Principle 1, reads "The Organization is based on the principle of the sovereign equality of all its Members" see <http://www.un.org/aboutun/charter/chapter1.htm>.

8 Annex 1 includes the 24 original OECD members, the European Union, and 14 countries with economies in transition. These countries have certain obligations within the UNFCCC and have also accepted emissions targets for the period 2008–12 as per Article 3 and Annex B of the Kyoto Protocol. These countries also have a special obligation to help developing countries with financial and technological resources. For more info see <http://unfccc.int/essential_background/ glossary/items/2639.php>.

countries go down dramatically. Agarwal and Narain's methodological intervention underscores that certain knowledge constructs of climate change "facts" are a key into the politics of distributing emissions limitations.

The per capita position was then taken up by the Indian delegation at the second meeting of the Intergovernmental Negotiating Committee (INC) in June 1991, in Geneva. The Indian proposal included a statement that emissions should "converge at a common per capita level" (INC 1991, 12). India's reason for articulation of the per capita position was to highlight the differences in responsibility for causing climate change. Their argument found its way into the convention as Article 3 of the UNFCCC, recognizing that states have "common but differentiated responsibilities".[9] Many developing countries (and one or two European countries) supported the Indian position, but, in the end, they lost the battle. In the final drafting stages of the framework convention, the principle of accounting for emissions on a per capita basis was put in brackets and it never appeared as an item in the convention text that was adopted on the 9th of May, 1992. However, echoes of the debate have been heard later in the climate negotiations.

The work towards the agreement that later became the Kyoto Protocol began when the first Conference of the Parties (COP1), in Berlin in 1995, adopted the "Berlin Mandate". This was a commitment to arrive at an agreement at COP3 in 1997 that would include "quantified, limitation, and reduction objectives" for Annex 1 parties, i.e. the industrial countries. Shortly thereafter, the Ad Hoc Group on the Berlin Mandate (AGBM) was initiated to provide the venue for debating and negotiating the burden-sharing scheme that would become the foundation of the Kyoto Protocol. In a review of a range of proposals submitted to the AGBM process, Ringius et al (2002, 14) note that proposals from France, Switzerland and the EU built on the idea of convergence of per capita emissions over time.

Another echo is to be found in 1995 in the controversy over economic dimensions in chapter 6 in the IPCC's second assessment report (SAR). The chapter was originally drafted as a comprehensive overview of climate change damages. Pearce's model (Pearce et al 1996) included differential values for lives in North and South to calculate global cost-benefit analysis of emissions abatement. CIESIN (1995) reports that early drafts of the chapter led to contentious debates at official presentations as well as on the ecological economics mailing list which included a formal protest endorsed by hundreds of scientists and researchers. The chapter was rejected at a meeting in Geneva and the key elements of the rewritten chapter were not included in the summary for policymakers (a key part of IPCC's assessment reports). Well articulated objections to the chapter were raised by Masood (1995) and by Meyer (1995) and his Global Commons Institute, who out of this developed

9 The whole article reads: "The Parties should protect the climate system for the benefit of present and future generations of humankind, on the basis of equity and in accordance with their common but differentiated responsibilities and respective capabilities. Accordingly, the developed country Parties should take the lead in combating climate change and the adverse effects thereof."

the notion of "contraction and convergence" (C & C). The C & C framework models how the trajectory of emissions would travel if we start from a status quo emissions distribution and move towards per capita equality (convergence) while reducing emissions to an overall level which is a politically set goal to achieve climate stability (contraction). It is interesting to note the range of actors who have expressed support for C & C.[10] They are able to do this precisely because the framework articulates concerns of the South for equal per capita emissions with US/Western concerns for "meaningful commitments" from developing countries.

In some sense, the "equal per capita" position poses a challenge to territoriality in that it relates individuals (as humans) to the atmosphere as a global common. It establishes individuals' equal access and responsibilities according to universal principles. However, those individual CO_2 contributions are actually calculated on the basis of national emissions and then divided by the number of people living in the state. Thus, in fact it is a territorially based "per capita position" that takes the emissions of the national community and divides by the population. Hence, the low Indian per capita level depends on the large amount of poor Indians with very low emissions, while the Indian middle class consumes carbon at or close to an OECD (Organisation for Economic Co-operation and Development) average. Thus the debates around justice, while being instinctively cosmopolitan and working in ethical terms alongside the scientific constructions of climate change as "global," in fact get drawn back to a statist account of rights. In the climate negotiations, the principle of "sovereign equality" is underpinning the agreed protocols for emission reductions. This is clearly reflected in the Kyoto Protocol where all Annex 1 parties agreed to reduce emissions by at least 5% from 1990 levels by 2008–2012. 1990 was decided upon as the "base year" and a "cap" was put on past emissions. Thus in the dominant version, it is Americans (or other OECD states' members) that have in effect an a priori "right" to emit CO_2 at the rate they do, and any collectively agreed proposals to limit emissions need to start from that rate of emission as the point of departure.

In the context of the climate negotiations, allocations based on past emissions is tellingly named "grandfathering." It is the grandfathers of a political community, who through past actions, have acquired a "status quo right" for the now living to continue to use the atmosphere in the way they want. This reflects the (now palpably patriarchal) master discourse of territorial sovereignty in two ways. First, the members of the political community in question are territorially defined and, second, they are justified in the claim that the community should accept any intervention in the way they choose to live their lives. The consequence of "equal per capita emissions," when the idea is put into concrete proposals, at least proposed by state negotiators, is that the scheme becomes mediated by the Southern state's right to negotiate on behalf of "its" people through the principle of non-intervention which means that the Indian state can conceivably argue for equal per capita emissions

10 See more at New Scientists coverage on the 10 of December 2003 <http://www.newscientist.com/ article.ns?id=dn4467> or at <http://www.gci.org.uk/>.

in international politics while rejecting a right of the international community to question the distribution of emissions within India.

The first meeting of the parties to the Kyoto Protocol was held in Montreal, Canada, in November 2005 (hence COP11/MOP1). In the aftermath of this meeting, negotiators decided on the next steps for the climate regime, including negotiations toward new binding commitments for the Kyoto's developed country parties and a softer commitment under the UNFCCC to open a dialogue on long-term cooperative action. The decision to consider further commitments for the Annex 1 countries beyond 2012 will certainly bring back questions of justice to the UNFCCC negotiations. The scholarly community is flooding the intellectual landscape with arguments about the best design of a post-2012 climate regime (Biermann and Brohm 2005; Hulme and Metz 2005; Kavuncu and Knabb 2005; Muller 2005; Pearce 2005). However, the problem, with most of those contributions is that they take the ethical issues inherent in the way international relations is organized for granted. One must realize that the fundamental question of "how we might live" within the context of world politics (Booth et al 2000) has "already" been given a crucial answer by IR (both the practice and the discipline). We know this answer fairly well. It is about the unspoken necessity and naturalness of living in territorially defined political communities. This is why international relations pose a certain challenge for reflecting on ethical considerations. The problem of inequality is already deeply inscribed in our modern accounts of the international (Walker 2002). As Franke states,

> … the traditions of international thought and practices are founded upon basic Western stories regarding how humans, necessarily, come to form political society within states and, furthermore, how these territorial communities themselves must necessarily engage one another in a dis-ordered "social" sphere (Franke 2000).

Walker and Franke highlight the important point that international relations cannot be understood as merely an empirical arena for making claims about what is equal and unequal because international relations already implies a specific framing of those issues. Therefore, the debate over the proper basis for counting emissions of carbon dioxide is clearly situated within a discourse concerning territory and territorialization. While the debate on how to allocate CO_2 emissions and future emissions cuts is often understood as the endpoints of a broad spectrum (state vs. per capita), one has to realize that the entire spectrum defines justice to be membership in a territorially based political community. Hence, carbon consumers over the world are territorialized rather than individualized.

These Sinks are Our Sinks

The fact that vegetation and soils absorb carbon has been well known for a long time, but in recent years the inclusion of this idea in the climate convention has been the subject of intense negotiations. The oceans and the terrestrial biosphere act as large "sinks" for carbon that can be released to the atmosphere through many different land-use practices. The Swedish scientist, Svante Arrhenius, understood already in the late nineteenth century that carbon flows throughout the Earth, and he made reference to the exchange of "carbonic acid" between three major carbon reservoirs: the atmosphere, the terrestrial systems and the oceans (Rodhe et al 1997). In the 1950s scientists started to construct carbon flux models that measured the amount of carbon stored and exchanged between the atmosphere, the ocean, and the terrestrial and geological reservoirs.

The natural state is that carbon flows in a global carbon cycle between the atmosphere, the oceans and the terrestrial biosphere and up until the mid 1990s carbon cycle science had been directed at measuring and depicting these global flows. When the UNFCCC was negotiated in the early 1990s there was a prevalent interest in the idea of "carbon sinks" and the convention encouraged its signatories to make inventories of its domestic sinks of GHG. One might say that the step from "making inventories" to "invent" is a short one. The invention of the national sink came to be solidified through a rapid and substantial shift in carbon cycle science that included agreement on the use of a common methodology for accounting for flows of carbon within the territory of the state.

Terrestrial carbon sinks entered the negotiating table in time for COP3 in Kyoto in 1997. Even though the idea of sinks had been around for quite a long time, "sinks only became the focus of intensive and high-level political debate in the closing stages of the protocol negotiations, as negotiators realized just how much was at stake" (Grubb et al 1999, 77). It was argued, for example by the US, that removal of atmospheric CO_2 by terrestrial ecosystems would not only offer the same climatic effect as emission reductions but also do so at significantly lower societal costs. Towards the end of the Kyoto negotiations, the United States and its allies in the so-called Umbrella Group (notably Canada, Japan, Russia and Norway) portrayed terrestrial carbon sinks as one of the keys to a cost-effective implementation of the Kyoto targets and a prerequisite for a final agreement (Grubb et al 1999). Article 3.3 in the Kyoto Protocol allows Annex 1 parties to account for removals by sinks "resulting from direct human-induced land-use change and forestry activities, limited to afforestation, reforestation and deforestation since 1990" (United Nations 1997). Article 3.4 also paves the way for an inclusion of sinks resulting from "additional human-induced activities" in a later stage of the negotiations.

The breakdown of the climate talks in The Hague had many reasons (Grubb and Yamin 2001; Ott 2001), but carbon sinks were clearly one of the most difficult issues to resolve (Dessai 2001; Sedjo et al 2001). COP6 was resumed in Bonn in 2001, and this time the negotiating parties were put under pressure to reach an agreement, as the new US administration led by President George W. Bush had announced that the US would not ratify the Kyoto Protocol. The Bonn compromise was built on a system of caps and discounts that would restrict the amounts of credits gained under Articles 3.3 and 3.4 and hence "manage" the large methodological uncertainty associated with the accounting of carbon accumulation in agricultural soils, grazing lands and harvested forests (Schultze et al 2002). The Bonn agreement was consolidated at

COP7 in Marrakech later that same year, and together these two meetings mark the end of the phase of negotiating a deal on sinks.

While carbon cycle science, until the signing of the climate convention in 1992, had been focused on carbon flows on a hemispheric or, at best, continental scale, the call for inventories of carbon sinks initiated new research to estimate terrestrial carbon uptake within state borders. The transformation of the global carbon cycle into territorial sinks is certainly not an anomaly in the way modern society relates to nature; it is rather a familiar one and indicative of modern practices of governmentality. As Kuehls put it: "the modern sovereign state is a particular political construction for which the environments do not come ready made. The task of molding environments to fit the sovereign state is that of government" (Kuehls 1998, 49). To capture carbon exchanges in various locations, many field sites have been established and an expanding grid of flux measurement stations has been installed in Europe, North America and in tropical regions. This growth in scientific research and measurement techniques, developed to meet political requirements to control and manage carbon flows within state boundaries, has reinforced the political representation of the "national carbon sink" and hence given the territorialization of global carbon flows continued legitimacy.

While it is easy to conceive of the sink issue as being produced by the master discourse on territorial sovereignty, the sink issue also illustrates a kind of rule over space that has more to do with the imperial than with the international. There are, in general, two sides of the imperial argument, which invoke both recent debates on the "new imperialism" in IR (e.g. Cox 2004) and debates emanating from Hardt and Negri's (2000) "Empire" thesis. Writings on the "new imperialism" are mostly about debating US power and the extent to which world order is imperial rather than hegemonic (e.g. Ikenberry 2004), while "Empire" is the claim that the world is indeed imperial, but that it is not an American empire but nevertheless a single logic of rule. Important for the interpretation of sinks here, is that Empire is best seen less in terms of a reterritorializing logic in the aftermath of September 11th, as a reorganizing of the deterritorializing logic of globalization through the combination of a single overarching set of rules to govern the global economy, and the application of military force to sustain that rule (see in particular Coward's 2005 interpretation of Hardt and Negri). The construction of sinks as nationally territorial spaces is nevertheless undertaken with the understanding that the management of such sinks serves a universal order.

The management of sinks through the Clean Development Mechanism in the Kyoto Protocol makes possible the reterritorialized control of the South by the North. However, the responsibility and authority over the "sink-spaces" (e.g. plantations and management of trees) are far from clear. In this sense, areas of carbon sinks bear a family resemblance with other cases of control over distant subordinate places, such as "debt-for-nature swaps" (Laferrière 1994), military bases (Johnson 2004), tourist resorts (Gössling 2002), economic processing zones (Abbott 1997) or by "supporting bioprospecting in biodiversity rich, postcolonial territories" (Eckersley 2004, 249). In this vein, Dalby notes that environmentalists "have frequently promoted the establishment of protected spaces, parks and the control of populations in manners that nonetheless replicate the practices of empire" (Dalby 2002, 8–9). Of course, this differs from earlier British and French territorializations in that the official politics is not imperial, but the similarities are still there.

In Fogel's (2004) analysis, the imperial argument acquires another twist. Fogel holds advisory scientists within the IPCC and other climate related institutions accountable for the idea of forests as "empty" and available space. She cites the US Department of Energy experts who have called for "the intensive management and/or manipulation of a significant fraction of the globe's biomass" (Fogel 2004, 110). Fogel sees the emerging culture of carbon management as contributing to a mechanistic "global gaze" that moves to standardize and enroll both people and the natural world into largely inaccessible global institutions. Fogel's view on the emerging sink discourse is reminiscent of the way that Hardt and Negri (2000) have captured governmentalities under the age of the fragmented, fluid and foundationless Empire. As Dalby summarizes the condition: "Sovereignty is bleeding away from states in some amorphous series of rules, regulations and shared procedures that exceed the mandates of states and set the terms for incorporation of many institutions and peoples into an amorphous but powerful arrangement they simply term 'Empire'" (Dalby 2002, 1–2).

An imperial rendering of the sink issue highlights that an emergent global culture of carbon management puts the world's entire biosphere under one type of rule. This "epistemological empire," while still being fragmented and fluid, has the potential to order the world so that the Earth's biosphere becomes one undivided territory.

This section has shown how the transformation of the global carbon cycle into national sinks is produced through a discourse on territoriality. Indeed, this transformation is not surprising as it fits into a historic lineage of Western political imagination that envisions the environment as part of the sovereign state system in the form of "natural resources." The sink issue also illustrates practices of environmental governance that might be called "imperial," that is to say, understood both as reterritorialized control over the South by the North and as the establishment of an epistemological empire of carbon management and control.

Conclusions: Contradictions of Reterritorializing Discourses

This chapter has investigated the interplay between constructions of the climate (as global) and world politics (as territorial). In the singing of climate politics into existence, many songs have been written about the spatial "mismatch" between the environmental (integrated) and the political (fragmented) in environmental thought and policymaking. It was a dominant lyric within the environmentalism of 1970s (for example in Ward and Dubos (1972) *Only One Earth*), and it was picked up in the official discourse when the environment became an issue within world politics. The opening phrase of the 1987 World Commission on Environment and Development stated that "The earth is one, but the world is not" and later on one could read, "From space, we see a small and fragile ball dominated not by human activity and edifice but by a pattern of clouds, oceans, greenery, and soils. Humanity's inability to fit its activities into that pattern is changing planetary systems fundamentally" (WCED 1987, 308).

We agree that such naïve appeals to a global "whole" that serves to transcend "narrow" nationalisms must be contested, as they precisely reproduce a naturalized account of political space just like the nationalisms they seek to replace. To return to the constructivist accounts of space discussed in the introduction, territory is itself a social construct; and thus, no appeal to particular types of space or spatiality as a foundation for political claims can be accepted. However, it has not been our intention in this chapter just to undermine those claims; rather, we want to emphasize that climate politics is situated between articulations of a debordered and rebordered political space. Processes of deterritorialization and reterritorialization form a major axis of complexity in climate politics.[11] The question is thus less "how can we get states beyond their parochial concerns?" and more "what precise form of discursive construction of territoriality and its alternatives permit and constrain the development of climate politics?" In terms of Smith's metaphor, "what sorts of songs might permit or obstruct what sorts of climate futures to emerge?"

The answer, emerging from the analyses above, is that the principal drivers of and constraints on climate policy from this perspective are twofold. First, there is the inevitable tension set up by the articulation of climate change as a "threat", a "danger" of an "urgent" nature. Environmentalists, scientists, and some politicians invoke this to justify global action, but it plays also into dominant accounts of security, which are nationalist and territorial. The space for articulating a question in terms of security has been narrowed to a nationalist agenda. Particularly in the US this has been found in the post-September 11 situation, but also more generally given the rise of neo-conservatism in the US, and its adoption of aggressive foreign policy positions since the end of the first Clinton administration, and with the establishment of the Project for a New American Century.

Mark Lacy (2005) has persuasively outlined the place of climate change in the traditional security imagination. Lacy engages specifically with realism's traditional

---

11 This chapter does not seek to engage "non-territorial" modes of climate and environmental governance as found for example in Pattberg 2005; Lipschtuz and Rowe 2005; Haas 2004; Shaw 2004; Bulkeley and Betsill 2003; Jagers and Stripple 2003. While most of these discussions sit fairly well within a traditional "pluralist" outlook on world politics, some are more geared towards disrupting the assumption of sovereign political spaces. However, what is emerging overall is not something that can be framed as an "alternative" (non-territorial) organization of environmental governance, but rather the insight that Agnew's (1994) territorial trap continues to identify the conceptual ground wherein which these debates are carried out. This is to say (in a very dense and blunt way) that because the norm (territorial) and the exception (non-territorial) implicate each other, a particular spatial duality is deeply ingrained in the conceptual architecture of global environmental governance (i.e. the private, the civil, the non-state, etc.) and stands in the way regarding future theoretical developments.

discrimination between first-order and second-order threats. His point is not so much that of those advocating a "widening" of the security agenda on the basis that "the environment is important too," but rather it is an investigation into the grounds that make it possible to sustain a hierarchy of threats. The result of this "hierarchy of security" is that climate change is regarded as, at best, a "second order concern" and, at worst, something which ought properly not to be considered a legitimate security concern at all: "You simply cannot be a Realist if you take non-traditional threats seriously; and to not be a Realist means that one can only be a utopian, a liberal, an idealist or radical. In short, you become someone who is not in touch with the 'reality' of global insecurity and uncertainty" (Lacy 2005, 31).

The second shaper of climate policy discourse in terms of territoriality is a set of already existing discourses about globalization, competitiveness, jobs, and economic growth. This narrows down the discourse of climate change as opportunity to one of an opportunity for *national* economies (with the exception of the EU which operates as a hybrid between a regional agreement and a territorial state). The opportunity consists of promoting particular patterns of investments to gain an advantage in emerging markets such as for renewable energy, emissions markets, efficiency technologies, and financing options.

The space that currently remains for articulating climate policy is dependent on the ability to draw on an ecological modernization discourse, which enables policy-makers and a sufficient bloc of important firms to move beyond a "climate-growth" dichotomy and to articulate this in terms of a national strategy, given prevailing discourses about globalization. In particular, a storyline that suggest that globalization creates opportunities for first-mover advantage and regulatory stimulation of growth. Much of the difference between the US and Europe can be situated along the lines of this discursive articulation. Steven Bernstein has summed up the implications for international environmental diplomacy:

> On one hand, the evolution of possible management regimes for global environmental problems should be expected to occur within the opportunities and constraints of liberal environmentalism. On the other hand, in cases where solutions to international environmental problems that fit within liberal environmentalism are evasive, international cooperation is likely to remain difficult (Bernstein 2002, 10).

This chapter has explored the constitutive aspect of territoriality in global climate politics. We wanted to focus, not so much on the question of "who constructs" (for example, scientists or policymakers) as on the categories that make certain constructions seem natural and tend to legitimize certain political economic solutions/ forces. Such a theme was captured some years ago by Ken Conca in his chapter on *Environmental Change and the Deep structure of World Politics*; "the ways we perceive and respond to ecological interdependence are likely to be structured along modern, sovereign, capitalist lines" (Conca 1993, 321), and he then suggests that future changes in environmental governance are likely to come from tensions within those "deep structures." Against the backdrop of a supposedly deterritorialized global climate, we have outlined three sites where the climate, and ensuing climate policy, is reterritorialized. While not always evident at immediate inspection, territoriality is present "behind the curtains" in environmental politics. It cast its shadow on the different articulations of "danger" and "opportunity" with regard to the climate issue, it returns in the debates around a fair and just distribution of burdens for emission reductions, and it is fundamentally there in the way global flows of carbon are articulated as "sinks" within the space of the state. Despite familiar claims to situate climate change within the fairytale of a world-without-borders, the climate issue illustrates the complexities inherent in the current reconfiguration of political space as a continuous tension of de- and re-territorial practices.

References

Abbott, J. (1997), Export Processing Zones and the Developing World. *Contemporary Review*, 270, pp. 232–238.

Agarwal, A. and Narain, S. (1998), "Global Warming in an Unequal World: A Case of Environmental Colonialism" in K. Conca and G.D. Dabelko, eds. *Green Planet Blues*. Boulder, Colorado: Westview Press, pp. 157–160.

Agnew, J. (1994), The Territorial Trap: The Geographical Assumptions of International Relations Theory. *Review of International Political Economy*, 1, pp. 53–80.

Alley, R.B., Marotzke, J., Nordhaus, W.D., Overpeck, J.T., Peteet, D.M., Pielke Jr, R.A., Pierrehumbert, R.T., Rhines, P.B., Stocker, T.F., Talley, L.D., and Wallace, J.M. (2003), Abrupt Climate Change. *Science*, 299, pp. 2005–2010.

Anand, R. (2003), *International Environmental Justice : A North-South Dimension*. Aldershot, Hants, England; Burlington, VT: Ashgate.

Arrhenius, S. (1896), On the Influence of Carbonic Acid in the Air Upon the Temperature of the Ground. *Philosophical Magazine*, 41, pp. 237–276.

Aspen Global Change Institute (2005), *Abrupt Climate Change: Mechanisms, Early Warning Signs, Impacts, and Economic Analyses 9–15 July*. Workshop summarized at <http://www.agci.org/abrupt.html#session%20description>.

Banuri, T., Goran-Maler, K., Grubb, M. Jacobson, H., and Yamin, F. (1996), "Equity and Social Considerations" in IPCC, *Climate Change 1995: Impacts, Adaptations and Mitigation of Climate Change: Scientific-Technical Analyses*. Cambridge: Cambridge University Press.

Barnett, J. and Adger, N. (2003), Climate Dangers and Atoll Countries. *Climatic Change*, 61, pp. 321–337.

Barry, J. and Paterson, M. (2004). Globalisation, Ecological Modernisation and New Labour. *Political Studies*, 52, pp. 767–784.

Bernstein, S. (2002), Liberal Environmentalism and Global Environmental Governance. *Global Environmental Politics*, 2, pp. 1–16.

Bieler, A. and Morton, A.D. (2001), The Gordian Knot of Agency-Structure in International Relations: A Neo-Gramscian Perspective. *European Journal of International Relations*, 7 (1), pp. 5–35.

Biermann, F. and Brohm, R. (2005), Implementing the Kyoto Protocol without the USA: The Strategic Role of Energy Tax Adjustments at the Border. *Climate Policy*, 4, pp. 289–302.

Booth, K., Dunne, T. and Cox, M. (2000), How Might We Live? Global Ethics in a New Century – Introduction. *Review of International Studies*, 26, pp. 1–28.

Boyle, S. and Ardill, J. (1989), *The Greenhouse Effect: A Practical Guide to the World's Changing Climate*. London: Hodder & Stoughton.

Broecker, W.S. (1987), Unpleasant Surprises in the Greenhouse. *Nature*, 328, pp. 123–126.

Brown, D.A. (2001), The Ethical Dimensions of Global Environmental Issues. *Daedalus*, 130, pp. 59–76.

Bulkeley, H., and Betsill, M.M. (2003), *Cities and Climate Change : Urban Sustainability and Global Environmental Governance*. New York: Routledge.

Bush, G.W. (2001), *Remarks by President Bush on Global Climate Change*. US Department of State. <http://www.state.gov/g/oes/rls/rm/4149.htm>.

Callendar, G.S. (1938), The artificial production of carbon dioxide and its influence on temperature. *Quarterly Journal of the Royal Meteorological Society*, 64, pp. 223–240.

Chatterjee, P. and Finger, M. (1994) *The Earth Brokers: power, politics and world development*. London: Routledge.

CIA (1974), *Potential Implications of Trends in World Population, Food Production, and Climate*. Central Intelligence Agency Report OPR-401, August.

CIESIN (Consortium for International Earth Science Information Network) (1995), *Thematic Guide to Integrated Assessment Modeling of Climate Change*. University Center Michigan. <http://sedac.ciesin.columbia.edu/mva/iamcc.tg/TGHP.html>.

Conca, K. (1993), "Environmental Change and the Deep Structure of World Politics" in R.D. Lipschutz and K. Conca, eds. *State and the Social Power in Global Environmental Politics*. New York: Columbia University Press, pp. 306–326.

Coward, M. (2005), The Globalisation of Enclosure: interrogating the geopolitics of empire. *Third World Quarterly*, 26 (6), pp. 855–571.

Cox, M. (2004), Forum on the American Empire. *Review of International Studies*, 30, pp. 583.

Dalby, S. (2002), *Environmental Security*. Minneapolis: University of Minnesota Press.

Dansgaard, W., Johnsen, S.J., Moller, J. and Langway, J.C.C. (1969), One Thousand Centuries of Climatic Record from Camp Century on the Greenland Ice Sheet. *Science*, 3 (166), pp. 377–381.

Dessai, S. (2001), Why Did the Hague Climate Conference Fail? *Environmental Politics*, 10, pp. 139–144.

Dessler, D. (1989), What's at Stake in the Agent-Structure Debate? *International Organization* 43, pp. 441–473.

Dobson, A. (2005), Globalisation, Cosmopolitanism and the Environment. *International Relations*, 19, pp. 259–273.

Doty, R. (1997), Aporia: A Critical Exploration of the Agent-Structure Problematique in International Relations. *European Journal of International Relations*, 3 (3), pp. 387–90.

Eckersley, R. (2004), *The Green State: Rethinking Democracy and Sovereignty*. Cambridge MA: MIT Press.

Fogel, C. (2004). "The Local, the Global, and the Kyoto Protocol" in S. Jasanoff and M.L. Martello, eds. *Earthly Politics: Local and Global in Environmental Governance*. Cambridge: MIT Press, pp. 103–126.

Franke, M. F. N. (2000), Refusing an Ethical Approach to World Politics in Favour of Political Ethics. *European Journal of International Relations*. 6, pp. 307–333.

Gössling, S. (2002), Global Environmental Consequences of Tourism. *Global Environmental Change*, 12, pp. 283–302.

Gribbin, J. (1990), *Hothouse Earth: The Greenhouse Effect and Gaia*. London: Bantam Press.

Grubb, M. (1995), Seeking fair weather: ethics and the international debate on climate change. *International Affairs*, 71 (3), pp. 463–496.

Grubb, M., C. Vrolijk, and Brack, D. (1999), *The Kyoto Protocol: A Guide and Assessment*. London: RIIA/Earthscan.

Grubb, M. and Yamin, F. (2001), Climatic Collapse at the Hague: What Happened, Why, and Where Do We Go from Here? *International Affairs*, 77, pp. 261–276.

Haas, P.M. (2004). Addressing the Global Governance Deficit. *Global Environmental Politics*, 4, pp. 1–16.

Hardt, M. and Negri, A. (2000), *Empire*. Cambridge MA: Harvard University Press.

Hargroves, K. and Smith, M.H., eds. (2005), *The Natural Advantage of Nations: Business Opportunities, Innovation and Governance in the 21st Century*. London: Earthscan.

Hay, C. (1995), "Structure and Agency" in D. Marsh and G. Stoker, eds. *Theory and Methods in Political Science*. Basingstroke: Macmillan Press, pp. 189–208.

Hulme, M. and Metz, B. (2005), Climate Policy Options Post-2012 – European Strategy, Technology and Adaptation after Kyoto – Preface. *Climate Policy*, 5, pp. 243–243.

Ikeme, J. (2003), Equity, Environmental Justice and Sustainability: Incomplete Approaches in Climate Change Politics. *Global Environmental Change-Human and Policy Dimensions*, 13, pp. 195–206.

Ikenberry, J.G. (2004), Illusions of Empire: Defining the New American Order.. *Foreign Affairs*, 83, pp. 144–154.

INC (1991), *Compilation of Possible Elements for a Framework Convention on Climate Change Submitted by Delegations*. UN Document A/AC.237/Misc.5/Add.2. June.

Jagers, S. and Stripple, J. (2003), Climate Governance Beyond the State. *Global Governance*, 9, pp. 385–399.

Johnson, C.A. (2004), *The Sorrows of Empire : Militarism, Secrecy, and the End of the Republic*. New York: Metropolitan Books.

Kavuncu, Y.O. and Knabb, S.D. (2005), Stabilizing Greenhouse Gas Emissions: Assessing the Intergenerational Costs and Benefits of the Kyoto Protocol. *Energy Economics*, 27, pp. 369–386.

Knight, D.B. (1982), Identity and Territory: Geographical Perspectives on Nationalism and Regionalism. *Annals of the Association of American Geographers*, 72, pp. 514–531.

Kuehls, T. (1998), "Between Sovereignty and the Environment: An Exploration of the Discourse of Government" in K. Litfin, ed. *The Greening of Sovereignty in World Politics*. Cambridge: MIT Press, pp. 31–53.

Lacy, M.J. (2005). *Security and Climate Change : International Relations and the Limits of Realism*. New York: Routledge.

Laferrière, E. (1994), Environmentalism and the Global Divide. *Environmental Politics* 3 (1), pp. 91–113.

Lanchbery, J. and Victor, D. (1995), "The Role of Science in the Global Climate Negotiations" in H.O. Bergesen, G. Parman, and Ø.B. Thommessen, eds. *Green Globe Yearbook O F International Co-Operation on Environment and Development 1995*. Oxford: Oxford University Press, pp. 29–39.

Lefebvre, H. (1991), *The Production of Space*. Oxford, OX, UK; Cambridge, Mass., USA: Blackwell.

Lipschutz, R.D. and Rowe, J.K. (2005), *Globalization, Governmentiality and Global Politics : Regulation for the Rest of Us?* New York, NY: Routledge.

Lorenz, E.N. (1963), Deterministic Nonperiodic Flow. *Journal of the Atmospheric Sciences*, 20, pp. 130–141.

Lövbrand, E. and Stripple, J. (2006), The Climate as Political Space: On the Territorialization of the Global Carbon Cycle. *Review of International Studies*, 32, pp. 217–235.

Lyman, F. (1990), *The Greenhouse Trap - What We're Doing to the Atmosphere and How We can Slow Global Warming*. Boston: Beacon Press.

Masood, E. (1995), Developing Countries Dispute Use of Figures on Climate Change Impact. *Nature*, 376, pp. 374–374.

Merchant, C. (1992), *Radical Ecology*. New York: Routledge.

Met Office (2005), *Avoiding Dangerous Climate Change*. <http://www.stabilisation2005.com/>.

Meyer, A. (1995), Economics of Climate Change. *Nature*, 378, pp. 433–433.

Muller, B. (2005), Quo Vadis, Kyoto? Pitfalls and Opportunities. *Climate Policy*, 5, pp. 463–467.

National Research Council (US). Committee on Abrupt Climate Change (2002), *Abrupt Climate Change : Inevitable Surprises*. Washington, D.C.: National Academy Press.

Newman, D. (2004), "The Resilience of Territorial Conflict in an Era of Globalization." Paper presented at the SGIR Conference 'Constructing World Orders' The Hague, September 9–11. (A revised version of the paper will be published in a forthcoming volume to be edited by Miles Kahler and Barbara Walter.)

Oppenheimer, M. and Boyle, R. (1990), *Dead Heat: The Race Against the Greenhouse Effect*. New York: Basic Books.

Ó Tuathail, G. (1996), *Critical Geopolitics: The Politics of Writing Global Space*. Minneapolis: University of Minnesota Press.

Ott, H.E. (2001), Climate Change: An Important Foreign Policy Issue. *International Affairs*, 77, pp. 277–296.

Paterson, M. (2001a), Climate Policy as Accumulation Strategy: The Failure of COP6 and Emerging Trends in Climate Politics. *Global Environmental Politics*, 1, pp. 10–17.

________. (2001b), "Principles of Justice in the Context of Climate Change" in U. Luterbacher and D.F. Sprinz, eds. *International Relations and Global Climate Change*. Cambridge MA: MIT Press, pp. 119–126.

________. (2001c). Risky Business: Insurance Companies in Global Warming Politics. *Global Environmental Politics*, 1 (4), pp. 18–42.

________. (1996), "International Justice and Global Warming" in B. Holden, ed. *The Ethical Dimensions of Global Change*. London: MacMillan, pp. 181–201.

Pattberg, P. (2005), The Institutionalization of Private Governance: How Business and Non-Profits Agree on Transnational Rules. *Governance – an International Journal of Policy, Administration and Institutions*, 18, pp. 589–610.

Pearce, D.W., W.R. Cline, A.N. Achanta, S. Fankhauser, R.K. Pachauri, R.S.J. Tol, and Vellinga, P. (1996), "The Social Costs of Climate Change: Greenhouse Damages and Benefits of Control" in J.P. Bruce, ed. *Climate change 1995 economic and social dimensions of climate change. Contribution of Working Group III to the Second Assessment Report of the Intergovernmental Panel on Climate Change*. Cambridge: Cambridge University Press.

Pearce, F. (2005), Tear up Kyoto or Make It Tougher? *New Scientist*, 186, pp. 12–13.

Porter, M. and van der Linde, C. (1995), Toward a New Conception of the Environment-Competitiveness Relationship. *Journal of Economic Perspectives*, 9 (4), pp. 97–118.

Ringius, L., Asbjörn, T. and Underdal, A. (2002), Burden Sharing and Fairness Principles in International Climate Policy. *International Environmental Agreements: Politics, Law and Economics*, 2, pp. 1–22.

Rodhe, H., Charlson, R. and Crawford, E. (1997), Svante Arrhenius and the Greenhouse Effect. *Ambio*, 26, pp. 2–5.

Rosa, L.P., and Munasinghe, M. (2003), *Ethics, Equity, and International Negotiations on Climate Change*. Cheltenham, UK; Northampton, MA: Edward Elgar Pub.

Ross, A. (1991), *Strange Weather: Culture, Science and Technology in the Age of Limits*. London: Verso.

Rowlands, I.H. (1997), International Fairness and Justice in Addressing Global Climate Change. *Environmental Politics*, 6, pp. 1–30.

Ruggie, J.G. (1993), Territoriality and Beyond – Problematizing Modernity in International-Relations. *International Organization*, 47, pp. 139–174.

Sachs, W. (1993), *Global Ecology. A New Arena of Political Conflict*. London and New Jersey: Zed Books.

Salinger, J. (2000), "Climate Change: Developing Southern Hemisphere Perspectives" in T.W. Giambelluca and A. Henderson-Sellers, eds. *Climatic Change*, 45, pp. 383–386.

Schultze, E.D., R. Valentino, and Sanz, M. J. (2002). The Long Way from Kyoto to Marrakesh: Implications of the Kyoto Protocol Negotiations for Global Ecology. *Global Change Biology*, 8, pp. 505–518.

Schwartz, P. and Randall, D. (2003), *An Abrupt Climate Change Scenario and Its Implications for United States National Security*. US Department of Defense, <http://www.environmentaldefense.org/documents/3566_AbruptClimateChange.pdf>.

Sedjo, R., Marland, G. and Fruit, K. (2001), Accounting for Sequestered Carbon: The Question of Permanence. *Environmental Science and Policy*, 4, pp. 259–268.

Shaw, K. (2004). The Global/Local Politics of the Great Bear Rainforest. *Environmental Politics*, 13, pp. 373–392.

Shiva, V. (1993). "The Greening of the Global Reach." in J. Brecher, J. Childs, and J. Cutler, eds. *Global Visions*. Boston: South End Press, pp. 53–60.

Shue, H. (1993), Subsistence Emissions and Luxury Emissions. *Law & Policy*, 15 (1), pp. 39–59.

________. (1992), "The Unavoidability of Justice", in A. Hurrel and B. Kingsbury, eds. *The International Politics of the Environment*. Oxford: Oxford University Press, 373–397.

SMIC (1971), *Report of the Study of Man's Impact on Climate*, the Royal Swedish Academy of Sciences, the Royal Swedish Academy of Engineering Sciences, and sponsored by the Massachusetts Institute of Technology.

Smith, S. (2004), Singing Our World into Existence: International Relations Theory and September 11. *International Studies Quarterly*, 48, pp. 499–515.

Soja, E.W. (1989), *Postmodern Geographies : The Reassertion of Space in Critical Social Theory*. London; New York: Verso.

Suganami, H. (1999), Agents, Structures, Narratives. *European Journal of International Relations*, 5, pp. 365–386.

Tickell, C. (1977), *Climatic Change and World Affairs*. Cambridge MA: Harvard Studies in International Affairs.

Tonn, B. (2003), An Equity First, Risk-Based Framework for Managing Global Climate Change. *Global Environmental Change*, 13, pp. 295–306.

Toth, F.L., ed. (1999), *Fair Weather? Equity Concerns in Climate Change*. London: Earthscan.

United Nations (1997), *Kyoto Protocol to the United Nations Framework Convention on Climate Change*. <http://unfccc.int/essential_background/kyoto_protocol/items/1678.php>.

________. (1992), *United Nations Framework Convention on Climate Change*. New York: United Nations.

Walker, R. (2002), International/Inequality. *International Studies Review*, 4, pp. 7–24.

Ward, B. and Dubos, R.J. (1972), *Only One Earth; the Care and Maintenance of a Small Planet*. New York: Norton.

WCED (1987), *Our Common Future – Report of the World Commission on Environment and Development*. Oxford: Oxford University Press.

Weart, S. (2003), *The Discovery of Global Warming*. Cambridge, MA: Harvard University Press.

Wendt, A. (1987), The Agent-Structure Problem in International Relations Theory. *International Organization*, 41, pp. 335–370.

Wiman, B.L.B. and Chong, S.M. (2000), "Introduction: From Climate Risk to Climate Security." in B.L.B. Wiman, J. Stripple, and S.M. Chong, eds. *From Climate Risk to Climate Security*. Stockholm: Swedish Environmental Protection Agency.

World Meteorological Organization. (1979),. *Proceedings of the First World Climate Conference-a Conference of Experts on Climate and Mankind*. WMO Publication No. 537.

Chapter 8

Trust Through Participation? Problems of Knowledge in Climate Decision Making

Myanna Lahsen

Introduction

Our discussion was lively and ranged widely when I in the late 1990s interviewed an influential US global change science administrator in a federal agency in Washington DC. At his initiative, we embarked on the issue of distrust in science and the related issue of participation in international forums under the United Nations Framework Convention on Climate Change (UNFCCC). He said:

> Given the suspicions that exist in the world—and which I have gotten in touch with better now but obviously still don't fully understand—imagine you're a South African [diplomat], and you're at a negotiation. And there are some scientists standing up there and they are saying: …'by the way, we've discovered through these measurements that South Africa is a major carbon source.' Now, if you're the leader of South Africa, are you happy that you are going to be responding to a monitoring program in which you have not a single investigator? No! It is unacceptable! It will not work. [*Small pause*] Now, if someone stands over there and says 'I am from such and such a project group, and we've done global monitoring, and it shows this for America and this for Europe and this for South Africa.' If the guy from South Africa knows that he had two investigators that were integral to that study and one of them is at MIT [Massachusetts Institute of Technology] getting a degree right now, suddenly those data have meaning. And it is literally the difference between assuming a conspiracy and assuming that the information is objective.

The science administrator articulates a dominant discourse of science according to which participation in the production or adjudication of scientific facts ensures that the latter will be viewed or described as such by scientists and decision makers. Scientists and science administrators are especially likely to reinforce this discourse which reflects assumptions at odds with constructivist understandings of science. As argued in other chapters in this volume as well, scientific facts, and hence also, of course, discourses about them, do not transcend particularities of perspective. If scientific interpretations are inextricably interwoven with politics and particularities of perspective, the fact of receiving an education abroad does more than merely enhance technical capacity of individuals: it also shapes subjectivities and political agendas. Integrating this insight, constructivist literature on the effectiveness of international cooperation around the environment identifies capacity building as a process that transforms values, beliefs, expectations and policy preferences (VanDeever 2005; Lahsen 2004; Mol 2002; Conca and Dabelko 2002; Cortell and Davis 2000).

While constructivist literature challenges dominant, objectivist discourses related to science, as a whole, it has attended insufficiently to the full range of political dimensions and consequences of such transformations, including problematic power dynamics whereby geopolitical and material advantage, such as that enjoyed by the United States, might translate into the ability to "preclude virtually any undesired normative developments, drown out competing frames, and ... attempt to shape potential outcomes according to [particular] instrumental interests" (Payne 2001, 53–4). There is an insufficient amount of empirically based literature investigating the extent to which power inequities impact scientific interpretations and associated political agendas, including the more diffuse effects of power, such as the full range of consequences of entraining less developed country scientists into international science, to return to the example above. To what extent does the fact that many scientists from less developed countries receive their educations in the US and Europe and participate in international science shape the problems on which they work and how they think of these problems? To the extent that it does affect their selection and construction of problems at the scientific level, what are the practical and political consequences? To what extent are their problem choices and constructions bundled up with particularities of normative structures and political agendas? The quote above begs these questions and an additional series of questions related to trust: To what extent do suspicions related to science exist and shape global environmental politics? How much is known of their systemic causes, including the role played by global inequalities in scientific capacity and power? To what extent does participation and scientific capacity reduce suspicions and their expressions and practical consequences?

It can be difficult to identify expressions of distrust related to science, especially when they are part of what one, following Michel Foucault (1980), might call a "subjugated" construction of science, a way of understanding that functionalist and systematizing thought suppresses and devalues as illegitimate, inappropriate, inferior and wanting at the levels of cognition or scientificity. On the other hand, one might argue that objectivist understandings of science that have dominated what Foucault refers to as functionalist and systematizing thought are no longer (as?) dominant. The strength of science as a force in the rhetoric of liberal-democratic politics has been eroded by new meta-narratives, and deconstruction of scientific knowledge has become an increasingly marked feature in policy related discourses (Lahsen forthcoming; Lahsen 2004; Fischer 2003; Marcus 1995; Beck 1992; Ezrahi 1990; Jasanoff 1990a). Nevertheless, objectivist discourses related to science arguably remain dominant. As I have shown elsewhere, deconstructions of science in political arenas also tend to be partial and "lop-sided," as actors typically deconstruct the scientific arguments of their opponents while resorting to objectivist language to promote their own preferred scientific interpretations and political agendas (Lahsen 1998).

In what follows, I reflect on the above-mentioned questions concerning trust on the basis of scholarly literature and empirical research among Brazilian environmental scientists and decision makers responsible for Brazil's foreign policy in the area of human-induced climate change. I discuss indications of distrust related to scientific knowledge underpinning international environmental negotiations, as evidenced especially on the part of less developed state leaders. I argue that the role of such intersubjective factors in climate change politics needs to be better understood, and relate this knowledge gap to broader tendencies in the fields of global environmental politics, international relations and beyond. The conclusion offers some thoughts about how to fill the gap, and draws out implications of the empirical data for the common emphasis on national scientific capacity and participation as a means of ensuring developing countries' trust in global environmental negotiations and associated science. It suggests that solutions to the problem of knowledge and distrust in international negotiations requires more deep-cutting solutions.

Below, I refer to non-discursive phenomena such as perceptions and economic structures. However, following a Gramscian framework, I understand economic dimensions to structure (but not determine) interpretive dimensions and, thus, discourses. Following others (Najam 2005; Williams 2005, 1993), I do not conceive of less developed countries (or the global "South") as a merely economic category but also as a political coalition associated with particular interpretive inclinations in international politics even as they do not share meaning in a uniform, monolithic nor unchanging manner. Consistent with this, while less developed countries' positions and discourses related to the global environment are heterogeneous, they have been remarkably consistent in their expressed aspirations and demands (ibid.).

I also diverge from a purely discursive analysis when referring to perceptions, and to distrust in particular. However, I acknowledge that it is impossible to access perceptions and experiences in any unmediated manner (Foucault 1980), and thus also to distinguish between suspicions and their expressions, whether verbal or non-verbal. To the extent that I here may seem to distinguish between perceptions and their expressions, I mean to indicate that dominant and official discourses related to science tend to omit discussion or expression of suspicions, and to suggest that some cognitive dimensions are thus suppressed, rendered more or less invisible. To the extent I argue that distrust exists, this is, nevertheless, on the basis of expressed manifestations, mostly discursive in nature, some instances of which are presented here.

The Role of Science in Global Environmental Politics and Its Treatment in International Relations

When strong assumptions and interests are at stake—whether rooted in shared disciplinary orientations, economic interests, or political convictions—uncertainties tend to be highlighted as actors seek to impugn the quality of countervailing science. This dynamic appears to be a general one, applying in the North (Lahsen 2005a, 1998; Oreskes 2004; Sarewitz 2000)[1] as in the South (Lahsen and Öberg 2006; Lahsen 2004; Lahsen 2001). The importance of science in international environmental regimes is thus also disputed. Backed by other, subsequent studies (Andresen, et al 2000), an analysis of international environmental treaties of the decades up until the early 1990s found science to play "a surprisingly small role in issue definition, fact-finding, bargaining, and regime strengthening" (Susskind 1994, 63). Yet, subsequent studies suggest that scientific input is critical to environmental policy formation (Dimitrov 2006; Haas 2004; Miles et al 2001; Mitchell et al 2005). At a minimum, scientific knowledge constitutes a necessary (albeit by no means sufficient) condition for policy advancement, shaping political discussions and outcomes as much as these shape competing framings of scientific knowledge. Moreover, the discussion that follows suggests that drawing conclusions about the overall role of science in international environmental regimes is premature, since much research has yet to be done to better identify its role.

Numerous studies stress the importance of the design and dynamics of the science-policy interface for scientific knowledge to impact environmental decision making processes (Mitchell et al 2005; Fogel 2004; Siebenhüner 2003; Cash et al 2003; Cash and Moser 2000; Miller 1998; Global Environmental Assessment Project 1997; Pielke Jr. 1994). Information use and effectiveness are known to depend on multiple factors, including how the information is distributed and the nature of decision makers' interpretive frameworks and political agendas (Stern and Easterling 1999; Jasanoff and Wynne 1998; Global Environmental Assessment Project 1997). Yet there is little consensus on how to bridge the gap between science and policy (McNie forthcoming; Smith and Kelly 2003). A comprehensive conceptualization of science-policy interfaces at the national and international levels is not easily forthcoming because of an inadequate amount of investigation into how knowledge systems work and how they might be better integrated with decision making to facilitate sustainability (Cash et al 2003; Bradshaw and Borchers 2000). The research gap reflects a more general lack of critical, empirical investigation at the nexus of science, technology and politics in general, and in environmental politics in particular (Jasanoff 2004, 1996).

The knowledge gaps related to knowledge systems and the global environment are particularly acute in the case of less developed countries (LDCs). Paul F. Steinberg has articulated the problem as follows:

> At present, environmental policymaking in developing countries is rarely studied and poorly understood. Social science research on global environmental problems has clustered at two levels of analysis—international cooperation and local resources management—

---

1 For an interesting study revealing the correlation between attributions of "junk science" and ideological bias, see Herrick and Jamieson 2000. For analysis of the varied intensity of political contestation of science in different national political cultures, see Jasanoff 1990b.

> leaving a gap where one would hope to find studies exploring the dynamics of national policy reform in the South [...] The result provides little guidance for understanding domestic-international linkages in the South, where most of the world's people, land, and species are found (Steinberg 2001, 5).

The role of science in developing states' environmental policy making receives even less attention.

The continued influence of the rational choice model in IR as in the social sciences more broadly (Rayner et al 2002) is one of the obstacles to filling the above knowledge gaps related to science, power, capacity, and participation. This model posits decision makers as strongly motivated to optimize integration of new information into their decisions, while sociological studies reveal the knowledge transmission process to be highly uneven, complex, difficult and varied depending on socio-cultural, institutional and political factors, including the characteristics of the receiver, the sources of the knowledge, and the type of knowledge at hand as well as its implications (Rayner et al 2002; Jasanoff and Wynne 1998; O'Riordan, Cooper and Jordan 1998; Proctor 1998; Sarachick and Shea 1997; Shapin 1995; Powell and DiMaggio 1991; Douglas and Wildavsky 1984). Typically focused on the actors who use science to mobilize around the same issue, the field pays less attention to areas where such mobilization has *not* occurred. Yet, as has been argued by analysts focused on regime formation (Dimitrov 2006) and norm transmission (Payne 2001; Checkel 1999), understanding the complex and possibly contradictory effects of science in international treaties requires similar examination of instances of failed epistemic convergence. The insufficient investigation related to science—and to knowledge more broadly—in international affairs characterizes various levels of analysis, from the implications of normative convergence or divergence through science to the differential effect of various types of knowledge. For instance, IR literature, including that on epistemic communities, has paid scant attention to problematic dimensions and limits of the supposed normative convergence often associated with global environmental problems and with science (Lahsen 2004, 2001; Miller 1998; Jasanoff 1996; Yearley 1996).

Radoslav Dimitrov (2006, 2003) argues that policy makers tend to act when they have reliable information of the socio-environmental consequences of any given environmental threat, and that this becomes apparent only when breaking knowledge down into different domains related, respectively, to the extent of a problem, the causes of the problem, and its consequences. Without denying the importance of studying the processes by which science is produced, legitimated, and accepted or rejected, Dimitrov leaves out of his analysis the interlinked issues of power, culture and reception, choosing to focus on knowledge only as an independent variable, that is to say, as a finished, legitimized product. While he justifies this by claiming that the dominant trend in IR is to treat the role of science in environmental policy processes as dependent on discursive strategies shaped by interests, values and power, recent literature reviews (Lahsen and Öberg 2006; McNie forthcoming) suggest that relatively little work has been done in IR and beyond to answer questions such as these: Why, and by what processes, do some scientific and political framings of issues come to be seen as reliable and authoritative? To what extent, and in what ways, do factors such as (lack of) capacity, trust, or dependence influence the shaping of knowledge and its acceptance or rejection? The scarcity of work answering such questions is a function of a number of factors, including (1) the general difficulty and disinclination in academic environments, and specifically among IR scholars, to study less tangible, intersubjective factors (Litfin 2000); (2) a general tendency to value "hard" and quantitative approaches over "soft" and more qualitative approaches; and (3), the dominance of the rational, unitary actor models and the associated limited impact of critical social theory in the field of IR. These tendencies work to place a "black box" around decision making processes, and to preclude normative questioning of the relationship between science (including the increasing role of scientific expertise) and democratic governance, including the impact of issues of trust, legitimacy and authority associated with the uptake of knowledge (Bäckstrand 2003; Litfin 2000; Jasanoff 1996).

The scarcity of work in this area persists despite calls for greater attention to power/knowledge dimensions in global environmental politics (Jasanoff and Martello 2004; Jasanoff 1996) and indications of the importance of such social dimensions, also referred to as "soft systems" or "social capital," in the transfer and uptake of knowledge (Smith and Kelly 2003; Putnam 1993).

Indications that Suspicions Related to Science are Important

In order to understand how issues related to culture and power impact global environmental politics, I return to my original questions: to what extent do suspicions related to science exist and shape global environmental politics? And to the extent that they do, are their systemic causes known, including the role played by global inequalities in scientific capacity and power?

The science administrator quoted in the introduction indicated the important role of suspicion related to scientific knowledge in international environmental politics. Emerging empirical studies support his observation and relate the so-called "North-South divide" that marks global environmental politics to inequities in national capacities to produce and frame knowledge and policy initiatives. Though the causes, dynamics and full range of consequences of the "North-South divide" remain insufficiently understood, there are indications that it reflects disenfranchisement on the part of LDCs tied to power differentials, including inequities in the area of human, technological, financial and informational resources (Fisher and Green 2004; Liverman and O'Brien 2002; Kandlikar and Sagar 1999).

Displaying broader societal tendencies in discourses related to science (Lahsen 2005a), it is still commonplace for IR scholars to characterize decision makers' attitudes to science as essentially trusting (Lahsen 2006) and marked by a perception of science as operating in a "rather rarefied atmosphere, immune to the vagaries of political power and subjective opinion … in a different realm and according to very different norms than politics" (Litfin 2000, 130). This may be what appearances often suggest, at least in international arenas. In domestic arenas, including the US Congress, suspicions and conspiracy charges abound, prompted by partisan politics and desires to prevent national support for the Kyoto Protocol and related policy agendas; rather than hidden, suspicions are used in US climate politics as political ammunition, and unwarranted charges of conspiracy are even deliberately produced when this serves powerful political interests (Lahsen 2005a; 1998).

In international arenas, by contrast, expressions of suspicion tend to be more muted, wherefore they are easily overlooked (Lahsen 2006). While suspicions also may serve powerful actors at the national level in LDCs (see below), in scientific and international arenas, they represent subjugated knowledge and are not readily expressed. An experience relayed by the US science administrator maker quoted above illustrates the need to probe beneath surfaces in order to identify commonly unspoken suspicions of the interplay of geopolitics and science on the part of LDC actors. This decision maker told me of an incident which impressed upon him the existence of distrust, even among collegial scientists and science administrators. In this case, scientists and administrators were working together to build and run the Inter-American Institute (IAI), a Brazil-based international organization supported by 19 countries in the Americas. As described on its website, the IAI is "dedicated to pursuing the principles of scientific excellence, international cooperation, and the open exchange of scientific information to increase the understanding of global change phenomena and their socio-economic implications" (IAI 2006). The science administrator was instrumental in creating the IAI and described the "startling" experience of realizing, some nine years into the project, that his Latin American collaborators suspected that it served to advance US geopolitical interests:

> I don't even remember what precipitated it but somehow something came along and a person from one of the countries of the Americas—from Chile—after 9 years of [being involved with] this, said "There it is! There is the US motive for IAI. I knew they were up to something, I knew there was a larger political motive. It took over eight years, but now it has been revealed." It was actually a group of people from several countries, joined by Chile, who said that IAI was an American rip-off. ... These are friends of mine, people I have known for years, and I suddenly realized: oh my God, they have been sitting there in their respective countries, these pals of mine, wondering what devious thing I was up to.

Another example is Joyeeta Gupta's empirical study which focused on the World Bank-coordinated Global Environmental Facility (GEF) and revealed suspicion and resentment on the part of less developed country representatives with regards to this institution which oversees funding for global change science and development projects in developing countries (Gupta 1995). Gupta found the GEF to be the object of deep, if generally unstated, suspicion and resentment among developing country representatives, who believed the GEF's institutionalized power hierarchy served developed (i.e., donor) states' interests. Her study suggests that suspicions about scientific knowledge extend to this multi-lateral institution, and the political processes and discussions that structure and surround it. As I describe below, my interviews in Brazil also revealed suspicions that the GEF uses scientific studies to obtain political effects. Most prominently, a UNFCCC-involved governmental decision maker described pressures from GEF and other multi-lateral organizations for Brazil to produce vulnerability and adaptation studies as an indirect attempt to weaken the government's ability to control climate-related political agendas at the national and international levels. Stressing the great uncertainties marring such impact and vulnerability assessments, he judged them unreliable as a basis for decision making but expected that they nevertheless would galvanize civil society against the government's strategy to emphasize mitigation over adaptation and place the burden for mitigation on developed countries.

Some argue that global resource disparities bearing on science and environmental policy shape such suspicions and the conditions that give rise to them. The dynamics of the science-policy interface in LDCs are different from those of developed countries in important respects because of resource disparities. Making this point, Milind Kandlikar and Ambuj Sagar (1999) identify five "gaps" that mark these disparities:

1. Resource gap: availability of human and material resources
2. Relevance gap: relevancy of existing research to issues faced by different countries and regions of the world
3. Participation gap: participation levels and input countries have in international scientific programs and processes
4. Perception gap: perceptions of the role and dynamics of research, analysis, and assessment processes—of what is being done, why, and how
5. Policy-culture gap: ability and approach to connect science and policy.

Northern nations, particularly the United States, overwhelmingly dominate the production and framing of science underpinning international environmental negotiations. An emblematic case in point is the Intergovernmental Panel on Climate Change which supports the UNFCCC. Less developed country scientists made up a total of 17.5 percent of the scientists producing and reviewing the IPCC's Third Assessment Report involved, with developed country scientists making up the difference with 82.5 percent (the figures are derived from Table 1 in Haas 2004, 582). In the production of the IPCC's *Special Report on Land Use, Land Use Change and Forestry (LULUCF)* in 2000, the US had roughly as many or more participating scientists as three continents combined (Africa, Asia and Latin America) while most peer reviewers originated in a handful of countries, notably the United States (Fogel 2004, 2002).

Analysts have paid scant attention to the existence and the policy consequences of such inequities in expert networks (Biermann 2000). According to Cathleen Fogel (2004, 2002), the unequal representation of LDC scientists in the politically consequential IPCC LULUCF report affected the policy outcome in favor of the more powerful developed countries, the United States in particular. In short, global inequity in states' abilities to produce science and direct research agendas has given rise to "an international climate change research enterprise that, when viewed from a Southern perspective, does not live up to its 'global' label;" an enterprise which, despite its apparent transnational dimensions, remains "headquartered in the North, comprised primarily of researchers in the North, dominated by Northern interests and agendas, and shaped by Northern perspectives" (Kandlikar and Sagar 1999, 133).

Less developed country representatives are not blind to their disadvantage in science-infused political discussions. A majority of IPCC-involved actors interviewed by Kandlikar and Sagar in the Indian context expressed that they, along with actors of the South as a whole, did not have much influence over the IPCC agenda (ibid., 134). Their inferior power reflects the more general economic weakness and associated "influence poverty" suffered by their countries, a shared condition among them that has engendered commonalities in their interpretive and discourse tendencies in global environmental politics (Najam 2005, 113). Frank Biermann's (2000) empirical studies in India similarly identified a perception of bias among actors there that the "international science" offered by transnational expert networks is biased and not to be accepted at face value. His interviews revealed "war[iness] of prejudice in the framing of assessments;" "great suspicion" of the IPCC and perceptions of it as "a 'political-scientific' institution with little transparency and inherent Northern intellectual supremacy" (Biermann 2001, 299). Cathleen Fogel's empirical studies of the production of the IPCC LULUCF report yielded expressions of disempowerment on the part of less developed country negotiators that harmonize with the aspirations and demands that have been voiced by these countries on environmental issues since the 1970s, in particular the desire for systematic change in global political relations (Najam 2005). Fogel perceived a "continuous and deep 'North-South' divide linger[ing] on palpably" in policy makers' engagement in forums related to the IPCC report in question, seeing in supposedly technical debates a microcosm of the mistrust and different perspectives between developed and developing countries on the issue of responsibility for climate change and the meaning of tropical deforestation (Fogel 2002, 366 and 267). Moreover, there are indications that at least some less developed country decision makers at times perceive science as "politics by other means" (Elzinga 1993) favoring dominant geopolitical powers. For instance, Fogel identified variation in the extent to which these global inequities in scientific capacity and power affected LDC delegates' reception of the IPCC LULUCF report. A number of delegates appeared to "perceive the report as relatively credible and non-controversial" (Fogel 2002, 338), whereas others described the report as a deeply political document. The latter delegates portrayed it as designed to advance hegemonic power, "a decoy mobilized by more powerful countries in the battle to prevent attention to the real issues at stake—developing country sovereignty and control of land" (ibid., 337–339).

Suspicions Related to Science in Brazil and the Promise of Participation as Solution

My own research in the Brazilian context confirms the above findings, including LDC policy makers' perceptions of the GEF and of links between science and domination. In interviews, Brazilian decision makers expressed suspicion that the GEF directs science agendas in LDCs in ways that favor Northern donor countries' policy preferences while weakening the Brazilian government's control over national climate affairs. Stressing the great uncertainties marring impact and vulnerability assessments, one decision maker judged them unreliable as a basis for decision making and resisted attempts by the GEF and other international organizations to get Brazil to produce such studies. He expected that the studies' uncertainties would be overlooked and that the international institutions pushing for the studies did so with the intention of galvanizing civil society to increase pressure for national policy action at odds with the government's two-pronged strategy to emphasize mitigation over adaptation and place the burden for mitigation on developed countries.

His prediction appears well-founded. Brazilian activist organizations, whose ability to mobilize civil society groups in the area of climate change has thus far been limited by the lack of detailed impacts studies (Lutes 2006), would likely use such studies to stimulate greater policy response at the national level. This example underscores the importance of scientific studies, or in this case, the deliberate absence of such studies, for the development of environmental policy and politics. It also suggests the ways in which suspicions related to science shape the dynamics of global environmental politics, but in ways that go largely uncharted in scholarly literature. As discussed below, an important part of the suspicions are tied to global disparities in power and scientific capacity.

Participation has been found to be of fundamental importance to the success of environmental assessments and associated international environmental policy initiatives. This was also evident in the statement by the US science administrator reproduced in the introduction. Prefaced with an acknowledgement of at best partial understanding of the nature and causes of the suspicions in global environmental science and related political arenas, he suggested a remedy to the attitudes of suspicion to environmental knowledge on the part of LDC decision makers: participation. As the IPCC's first chairman claimed in the beginning stages of the IPCC, many countries, and especially developing countries, "simply do not trust assessments in which their scientists and policymakers have not participated" (quoted in Siebenhüner 2003, 124). IPCC architects and policy analysts thus rightly stress the importance of national scientific participation and capacity for national political leaders' trust and involvement in the associated political negotiations (Mitchell et al 2005; Lahsen 2004, 2001; Siebenhüner 2003; Fogel 2002; Biermann 2002, 2000; Miller 1998; Global Environmental Assessment Project 1997).

Highlighting the importance of scientific capacity, a central Brazilian decision maker posited Brazilian capacity in the area of climate modeling as a prerequisite to the production of adequate national impact assessments. He expressed discomfort at having to rely on foreign climate model results, whose representation of climate-related systems in the Southern hemisphere he described as inadequate. The same decision maker also emphasized the need for better identification of baselines against which to produce estimates of future impacts of climate changes, and the difficulty of improving knowledge of baselines due to limited capacity. This suggests that resource disparities reduce the effectiveness of international efforts to assess and combat human-induced climate change.

Analysts also stress the importance of national scientific capacity to secure national interests in international environmental arenas:

> [I]ndigenous capacity to gather and analyze data, to build ones' own appropriate models, and 'deconstruct' those built by others is key to appropriately shaping international discussions and safeguarding national interests. Building internal capacity for knowledge generation and analysis in the South will be the first step in truly globalizing the climate discussions and feeding a variety of perspectives into the analytical efforts that are the basis of most policy considerations (Kandlikar and Sagar 1999, 135).

The above mentioned case studies (Siebenhüner 2003; Fogel 2002; Biermann 2001; Kandlikar and Sagar 1999) suggest that poorer states' limited scientific capacity, and associated weak participation and influence in political and scientific processes under the UNFCCC, leave their scientists and political leaders alienated and less inclined to trust the reports and the alleged concerns propelling them. These studies underscore the fundamental need to attend to how institutions such as the IPCC and UNFCCC are perceived, and to study the consequences and the structural causes of inequities in scientific capacity, representation and influence within these forums.

However, participation in itself is unlikely to solve the problem of distrust in international environmental politics. Importantly, since the problem of participation for the South is closely linked to disparities in scientific capacity, solving the problem of participation requires changes in the conditions causing the scientific disparities. Yet the recent emphasis on participation has not significantly altered these conditions, leaving unchanged the basic discursive and structural dimensions that empower richer states over poorer states in science-related processes such as those associated with the IPCC and the UNFCCC. This is true in this specific case and is also a critique more generally advanced against policy and development efforts emphasizing participation (Cleaver 2001). Moreover, everything else being equal, one cannot assume that improvements in participation and scientific capacity will necessarily reduce negative impressions and mistrust. Contrary to assumptions in IR and policy literatures, the outcomes of enhanced capacity and participation are unpredictable; the latter may in some instances invite rather than discourage distrust and disagreement (Jasanoff and Wynne 1998). Assumptions that transparency and participation enhance trust are premised on the faith that the processes guiding science are fair and will be perceived as such.

The US science administrator quoted in the introduction also manifests the above assumption. He takes for granted that, given the right credentials, the credibility of LDC scientists will be established in the eyes of their national political leaders. He assumes that the very act of participating in international science ensures recognition of the science in question as objective, and that national scientific participation renders politically consequential scientific findings trustworthy in the eyes of South African politicians, especially if the national scientists in question have received their educations from prestigious institutions in the global North, such as the prestigious MIT. In this hypothetical example, the South African scientists serve as expert witnesses who certify the inter-subjective nature of international science. Participation, in this model, is little more than a mechanism by which to reassure unnecessarily distrusting political leaders that everything is just and true. And it is assumed that if science is just and true (itself a construct, of course, albeit not recognized as such in this scenario), it will also be perceived as such. Much as it is impossible to observe climate change 'as it is,' perceptual filters shape understandings of science, especially in politically consequential issue areas involving significant uncertainties. In practice, South African scientists might interpret the scientific issues quite differently and perceive political bias in associated framings. And if they do not, this would not necessarily prove the objective truth of the scientific matter at hand, nor would it necessarily reassure South African decision makers. It could be that personal or political factors led the scientists to sign on to particular interpretations. It could be that the scientists choose not to express their disagreement, intimidated or resigned in the face of well-documented discrimination in science (Wennerås and Wold 1997; Gibbs 1995) and the intimidation less developed country participants can experience in international scientific assessment processes (Lahsen 2004; 2001). It also could be the case that scientists' educations in the North and their participation in Northern-dominated science had shaped their subjectivities and political agendas such that they accepted the science and associated framings, and/or that decision makers suspected this to be the case, with the effect of undermining their trust in national scientists. Aside from its potential to generate beneficial consequences, participation also can be a means of co-optation and control, aided by factors such as intimidation, the Abilene paradox, other forms of group-think and coercive persuasion (Cooke 2001). Likewise, apparent consent can be ambiguous and superficial, concealing dissent and ambivalence.

Such possible power-laden and ethical dimensions of normative convergence are often overlooked in IR (Lahsen 2004; 2001; Miller 1998; Jasanoff 1996). As a whole, development studies are similarly marked by a lack of critical reflective examination of deeper determinants of social change and policy processes in particular, including the impact of power-laden social relationships, social psychological processes, and access to and control over, information and other resources (Cooke and Kothari 2001).

Brazilian policy makers whom I interviewed (some of whom have themselves participated as both experts and policy makers in the production of the scientific assessments under the IPCC) did not reveal understandings of the associated processes and products as objective and apolitical. One policy maker, who is involved with the IPCC and the international negotiations related to climate change, thus described the IPCC as biased in favor of Northern framings of responsibility, a function of the huge disparity in scientific representation between more and less developed states. He claimed that the overwhelming representation of Northern scientists in the IPCC compared to their less developed counterparts, on the order of ten to one at the time of the 1999 interview, meant that the IPCC was "bound to reflect their perspective." Brazilian leaders thus believe that their lesser scientific capacity places them at a disadvantage in international science, and that this has political consequences. They perceive that their lesser scientific strength helps restrain their already more limited political power in the face of states with greater means and powers, rendering them comparatively less able to advance their perceived national interest in international forums (see Lahsen 2004; 2001). Thus, the policy maker mentioned immediately above criticized the IPCC for "playing dirty" by using its scientific edge to forge political advantage, suggesting that it advanced Northern political interests under false pretenses of value neutrality and objectivity (Lahsen 2004, 162–3). Studies focusing strictly on policy outcomes at the expense of the forces shaping the construction, deliberation and reception of associated knowledge and knowledge framings, fail to acknowledge this level of meaning-making and politics. As such, they can tend to encourage misleading perceptions regarding the existence of consensus and faith in science and in the processes by which science is produced and harnessed to political agendas (Lahsen and Öberg 2006).

Likewise, efforts at capacity training are generally understood by those designing these programs in uncritical terms, reflecting an inclination to conceive of science as unmarked by discourses and associated biases or, otherwise put, as little more than "a set of facts, skills, hypotheses, theories, and other information that can be communicated without reference to the social contexts of production, validation, or use" (Miller 1998, 11).

My research among Brazilian environmental scientists and decision makers suggests that the possibility of socio-political consequences of capacity building does not go unnoticed in Brazil, with roughly half of the persons obtaining PhDs presently doing so in the US (Ministry of Science and Technology 2006). To the extent that this possibility is perceived, the associated normative convergence is not uniformly or automatically assumed to be benign. This is true for decision makers with central power over Brazilian climate affairs. In interviews with me, decision makers within the Ministry of Science and Technology and Foreign Relations have described Brazilian scientists' ability to perceive the national interest as impaired by cultural and political indoctrination which accompanied their scientific training abroad and their engagement with international science in general. In one of the strongest expressions to this effect, a UNFCCC-involved policy maker in the Ministry of Science and Technology suggested that Brazilian scientists' foreign educations reduce their critical awareness and their ability to understand and serve national interests:

> If you don't have a kind of domestic way of thinking, that reflects in your thinking of [the environment and related policy issues]. You are like a parrot, you are repeating what

> people are teaching you. And even in universities you see this; people repeat what they hear in the literature. And that [literature] is from the developed countries.

"It is not a conspiracy," he added, it is that how you think reflects "the common sense of the community in which you live." I have encountered discourses to the same effect among other important decision makers in these two ministries. They understand international science as situated knowledge and a potential vector for hegemonic power.

At least some Brazilian scientists reflect on these questions themselves, acknowledging the possibility that their foreign educations and connections may bring them to align themselves with scientific and political agendas in conflict with more local and national environmental agendas and priorities (Lahsen 2004). However, decision makers' renditions of national scientists along these lines are also resented by scientists and associated with a top-down approach to decision making with deep roots in Brazilian political culture.

A subtle but real effect of this conceptualization of national scientists is to legitimate limited inclusion of national scientists and other segments of civil society in decision making related to politically charged environmental issues such as climate change and deforestation (Lahsen 2004). In other words, when Brazilian decision makers present the science as a hegemonic weapon in international politics, this can bolster tendencies in Brazil towards centralized decision making and the perpetuation of environmentally destructive development.[2] On the other hand, alternative tendencies to portray science as shaped by geopolitics are present as well and validated by extensive empirical studies. Thus, decision makers' perceptions to this effect may also lead them to counter prominent and potentially hegemonic renditions of science, as well as currents in Brazilian environmental politics favoring Northern problem framing agendas (Tesh and Paes-Machado 2004).

The extent of these perceptions among Brazilian decision makers is an open question requiring greater study; they are presented here mainly as a means of highlighting overlooked dimensions in IR and development literature related to science and policy, and the role of trust, participation and capacity building in particular. The extent to which Brazilian decision makers' perceptions of the relationship between science, power and politics have shaped the country's official position on climate change in international negotiations is unclear. Thus far, national trust in the science has not been put to great test; the "faults" in the science underpinning concern about human-induced climate change—in particular the general circulation models—have not yet become a focus at the national level. An important, likely reason is that Brazil stands to gain more from the UNFCCC in financial terms than it stands to lose, and, as commonly recognized, scientific evidence tends to be deconstructed to the extent that it threatens powerful political and economic interests (Lahsen 2005a; Jasanoff 1994; 1990a; Beck 1992; Dickson 1989).

---

2 Analyses suggest that authoritarian regimes as a whole tend to sacrifice the environment in favor of other concerns (Desai 1998).

Brazil has used the climate change negotiations to pursue long-standing national interests, even when they diverge from the position of the Group of 77 (G77) and China (Johnson 2001). In this, it is little different from the rest of the developing world. Despite pressures from Annex I states (in particular the United States), Brazil and China have led LDCs in their resistance to binding commitments under the Protocol. Argentina proposed voluntary commitments on the part of developing countries to reduce greenhouse gas emissions, but this idea was forcefully—and successfully—rejected by the Brazilian delegation, along with other G77 members and China. As informal leader of the G-77 coalition of LDCs with China in the climate negotiations, the Brazilian government has consistently upheld the Berlin Mandate according to which developing states need to reduce greenhouse gas emissions only after developed states have done so (Johnson 2001).

To the extent that developing states have shown interest in climate change negotiations, which they have done only to a limited extent (Najam, Huq and Sokona 2003), they have insisted on the principle of common but differentiated responsibilities. They have pushed strongly to reconcile national economic interests with environmental policy agendas, hinging their voluntary participation on Northern provision of financial resources and technology and insisting upon the need for capacity building and a longer time frame for the implementation of new rules in developing countries. The UNFCCC and subsequent international environmental policy agendas articulate and inscribe these demands (Williams 2005).

In short, free of commitments to limit economic interests in order to partake in the Kyoto Protocol, Brazilian leaders have not had compelling reasons for questioning the science or for voicing suspicions of the sort discussed above. The extent to which they will do so in the future is thus likely to depend on whether Brazil will be pressed to deepen its commitment under the Kyoto Protocol or adopt similar mechanisms in the future that could threaten powerful economic and political interests. As states begin discussions for the second commitment period, it appears highly uncertain whether Brazil (or by extension, the whole G-77 coalition) will even accept *voluntary* commitments for the second commitment period. The issue of distrust in science and associated global inequities in geopolitical and economic power are likely to impact a variety of issue areas requiring environmental decision making. For this reason, but also for ethical reasons beyond it, it behooves us to pay greater attention to the existence, dynamics and causes of the interplay between science and (dis)trust.

Conclusion

As this chapter has demonstrated, the implicit and explicit value of science in policy decisions is fundamentally shaped by socio-cultural dynamics at the level of systems, institutions, groups and individuals; yet, these dimensions receive relatively little attention in IR literature on climate change and global environmental politics in general. Even constructivist IR studies fall short when it comes to empirically grounded analysis of the mechanisms, dynamics and consequences of the diffusion of science and associated norms and interpretive frames.

Returning to the policy maker quoted in the beginning with the benefit of knowing the above-mentioned Brazilian understandings of international science and the role of foreign educations in advancing hegemonic agendas, it becomes clear that an MIT education and participation in international science is a double-edged sword which at times might *undermine* rather than *enhance* the credibility of LDC scientists in the eyes of their national decision makers. To the extent that future research reveals other LDC decision makers holding similar reservations regarding science to those in Brazil, this would increase the urgency of efforts to identify ways for LDC actors, not only government representatives but also relevant members of the scientific community and civil society, to participate on a more equal footing in science-laden international negotiations and associated processes.

The full range of consequences of science and the role of national scientists on the part of decision makers, in Brazil and beyond, has yet to be subjected to more extensive and systematic study sensitive to the particularities of national contexts. For instance, conceptualizations of science as hegemonic might at times be used to discredit and reject science when the latter supports inconvenient policy agendas. In addition, the locus of moral authority is a research question in need of reflexive, normative deliberation rather than something that can be decided a priori (Lahsen, forthcoming). There is a need to study perceptions of science and their impacts on a case-by-case basis and in cross-national perspective to identify global patterns and variations in understandings of the relationship between science and politics, as well as the structures that shape both. For instance, future studies might probe the factors explaining why LDC decision makers interviewed by Fogel differed in their expressions of relative trust or distrust with regards to the final LULUCF report. Did the differences reflect diverse interests in the findings and associated policy outcomes on this particular issue, and/or did they reflect deeper differences in political culture and in socio-political and economic structures shaping the science-policy interface in the various countries? Answers to these questions would be useful to the practical goal of improving international environmental policy and would help improve understanding of the interplay between science, power, culture and politics in a context of global environmental change. It would be interesting to know, for instance, whether LDC actors are less inclined to accept objectivist understandings of science compared to those from more developed and hegemonic countries, and if so, why and to what consequence, at the levels of theory *and* practice.

Also interesting to investigate is whether any given set of patterns related to perceptions of science shifts depending on the issue at hand. In the case of Brazil, do Brazilian decision makers only indicate anti-hegemonic understandings of science when this is politically expedient, while expressing more trusting and objectivist views when that is more conducive to political goals? Certainly, if taken at face value, Brazilian decision makers seem to embrace hegemonic science in their emphasis on the need for improved national scientific capacity in the area of computer modeling, in addition to considerable financial investments in that area, indicating that their skepticism regarding science and the reliability of national scientists is moderate. It bears emphasis that the evidence presented above was specifically selected to critique tendencies in IR and development literature, and does not capture the complexities characterizing Brazilian decision makers' attitudes towards science. Indeed, science is frequently invoked by Brazilians as an important weapon in defense of national territorial integrity and national interests (Lahsen 2005b). More definitive conclusions in the case of Brazil and beyond require additional analysis of discourses, uses, and policies related to science.

The central conclusion is that solutions to the problem of knowledge construction in climate change politics need to press beyond the common emphasis on participation and, even, enhanced national scientific capacity in LDCs. These common solutions are premised on an overly simplistic perception of the relationship between scientists and policy makers, obscuring the impact of globalization and the extent to which *lack of trust* characterizes relationships between decision makers and scientists at the national level. Such lack of trust, the above suggests, may be especially likely to characterize the science-policy interface in countries with limited means to direct national science agendas because of the disparities identified by Kandlikar and Sagar (1999). As noted, these disparities result from, and perpetuate, a general dependence on foreign donors and the fact that national scientists are acculturated in Northern science and associated interpretive frameworks.

Whatever else they do, it is clear that suspicions such as those identified among Brazilian decision makers above pose a fundamental challenge to assumptions that national participation—or variants of the idea, such as regional "centers of scientific excellence" in the global South (Huntingford and Gash 2005) will, in themselves, entirely solve the problem of how to produce and legitimize "global knowledge" related to climate change and the environment.

In sum, this analysis supports Stacy VanDeveer's (2005) point that analyses and programs intended to improve capacities bearing on environmental decision making tend to focus on the policy implementation phase, and that there is a need to attend relatively more to social processes, e.g., research, assessment and reception capacities and processes, that precede the implementation phase. Huge amounts of time and resources continue to be devoted overwhelmingly to the production and scientific assessment of climate science, while comparatively little attention is being given to whether or not intended audiences are receptive to the information being produced and, if they are, why, i.e., what factors have conditioned their attitudes, and with what practical and normative consequences?

If science is resisted and deconstructed as a function of its bearing on perceived interests, and if distrust in science indeed is prevalent among less developed country decision makers, the latter is likely to surface and shape global environmental politics more if these countries encounter mounting pressures to make binding commitments under the Kyoto Protocol, as is presently the case. It would thus benefit international policy efforts to attend to distrust related to science and to its causes, whatever their basis. Scholarly analyses can help by identifying national level interpretive biases and evaluating them in terms of democratic norms, ethics and equity. This would help identify obvious political interests underpinning various constructions of science as an institution and of scientific facts bearing on environmental policy, and help evaluate competing interests and views.

For practitioners, a first step might be to facilitate reflection and discussion among scientists and politicians about the directions of science agendas globally and about actual and potential political uses of science, along with associated normative and equity-related questions. Heeding the insight that science in many cases *is* the politics of climate change, and that the design and management of science agendas is a central medium through which social and political systems are produced and maintained (Jasanoff 2004, 1996; Lahsen 2001; Miller and Edwards 2001), such discussion must subsume open and broadly participative international debate about how to ensure greater participation by LDC actors not only in policy processes but also in shaping associated science agendas and problem framings worldwide. This is no tall order, but it is a more honest approach than the present maintenance of misleading non-constructivist constructions of science (Lahsen 2006), whether upheld by mere inattention or by the questionable assumption that a sanitized (but misleading) image of science is necessary for policy advance, and that this image, in fact, is being believed.

Doing away with objectivist discourses related to science can help reduce alienation and suspicion on the part of LDC actors skeptical that science is an apolitical and benign force in international politics (Lahsen 2006, 2004, 2001; Yearley 1996). Similar to Heather Smith's story of the Inuit hunter's intervention (this volume), bringing overlooked places, spaces and perspectives to bear on science agendas will reveal the Northern domination of "international" climate science and associated policy programs. While the increased complexity resulting from the inclusion of more and different voices challenges management, it promises to ultimately translate into policy advancement, not the least in light of the finding that states are more inclined to comply with international policy norms if they perceive themselves as a legitimate member of the international community of states. To the extent that the proposed science-focused discussions link policy advancement to resource and development concerns in LDCs, the result could be an instance of the civic environmentalism that Karin Bäckstrand and Eva Lövbrand (this volume) posit as the new discursive compromise between the global North and South in post Kyoto climate politics. It might help refresh thinking, and thereby also action, related to climate science and policy, though an obstacle to be expected will be the resilience of national and regional boundaries and global structures of inequity when it comes to funding, and thus, defining science agendas.[3]

3 This article was made possible by funding from the National Science Foundation (Grant No. 0242042) and the CLIPORE research program under the Mistra Foundation for Strategic Environmental Research through the Climate Science and Policy Beyond 2012 project. Any opinions, findings, conclusions or recommendations expressed in this material are those of the authors and do not necessarily reflect the views of the funding sources.

References

Andresen, S., Skodvin, T., Underdal, A. and Wettestad, J. (2000), *Science and Politics in International Environmental Regimes: Between Integrity and Involvement*. Manchester and New York: Manchester University Press.

Bäckstrand, K. (2003), Civic Science for Sustainability: Reframing the Role of Experts, Policy-Makers and Citizens in Environmental Governance. *Global Environmental Politics*, 3 (4), pp. 24–41.

Beck, U. (1992), *The Risk Society: Towards a New Modernity*. Newbury Park: Sage.

Biermann, F. (2002), Institutions for Scientific Advice: Global Environmental Assessments and Their Influence in Developing Countries. *Global Governance*, 8 (2), pp. 195–219.

________. (2001), Big Science, Small Impacts – in the South? The Influence of Global Environmental Assessments on Expert Communities in India. *Global Environmental Change*, 11, pp. 297–309.

________. (2000), *Science as Power in International Environmental Negotiations: Global Environmental Assessments Between North and South*. Belfer Center for Science and International Affairs (BCSIA) Discussion Paper 2000–17, Environment and Natural Resources Program. <http://Environment.Harvard.Edu/Gea>.

Bradshaw, G. A. and Borchers, J. G. (2000), Uncertainty as Information: Narrowing the Science-Policy Gap. *Conservation Ecology*, 4, p. 7.

Cash, D.W., Clark, W.C., Alcock, F., Dickson, N.M., Eckley, N., Guston, D.H., Jäger, J., and Mitchell, R.B. (2003), Knowledge Systems for Sustainable Development. *Proceedings of the National Academy of Sciences*, 100(14), pp. 8086–8091.

Cash, D.W. and Moser, S. (2000), Linking Global and Local Scales: Designing Dynamic Assessment and Management Processes. *Global Environmental Change-Human and Policy Dimensions*, 10 (2), pp. 109–120.

Checkel, J.T. (1999), Norms, Institutions, and National Identity in Contemporary Europe. *International Studies Quarterly*, 43, pp. 83–144.

Cleaver, F. (2001), “Institutions, Agency and the Limitations of Participatory Approaches to Development” in B. Cook and U. Kothari, eds. *Participation: The New Tyranny?* New York: Zed Books, pp. 36–55.

Conca, K. and Dabelko, G. (2001), *Environmental Peacemaking*. Baltimore, MD: Johns Hopkins University Press.

Cooke, B. (2001), “The Social Psychological Limits of Participation?” in B. Cook and U. Kothari, eds. *Participation: The New Tyranny?* New York: Zed Books, pp. 102–121.

Cooke, B. and Kothari, U., eds. (2001), *Participation: The New Tyranny?* New York: Zed Books.

Cortell, A.P. and Davis, J.W. (2000), Understanding the Domestic Impact of Norms: A Research Agenda. *International Studies Review*, 2 (1), pp. 65–87.

Desai, U. (1998), "Environment, Economic Growth, and Government in Developing Countries" in E. Desai, ed. *Ecological Policy and Politics in Developing Countries: Economic Growth, Democracy, and Environment*. Albany: State University of New York Press, pp. 1–46.

Dickson, D. (1989), *The New Politics of Science*. 2nd, Chicago: The University of Chicago Press.

Dimitrov, R.S. (2006), *Science and International Environmental Policy: Regimes and Nonregimes in Global Governance*. New York: Rowman & Littlefield Publishers, Inc.

________. (2003), Knowledge, Power, and Interests in Environmental Regime Formation. *International Studies Quarterly*, 47, pp. 123–150.

Douglas, M. and Wildavsky, A. (1984), *Risk and Culture*. Berkeley: University of California Press.

Elzinga, A. (1993), "Science as the Continuation of Politics by Other Means" in T. Brante, S. Fuller and W. Lynch, eds. *Controversial Science: From Content to Contention*. Albany: State University of New York, pp. 127–152.

Ezrahi, Y. (1990), *The descent of Icarus: Science and the transformation of contemporary democracy*. Cambridge, Mass.: Harvard University Press.

Fischer, M.M.J. (2004), *Emergent forms of life and the anthropological voice*. Durham, NC: Duke University Press.

Fisher, D.R. and Green, J.F. (2004), Understanding Disenfranchisement: Civil Society and Developing Countries' Influence and Participation in Global Governance for Sustainable Development. *Global Environmental Politics*, 4 (3), pp. 65–84.

Fogel, C. (2005), Biotic Carbon Sequestration and the Kyoto Protocol: The Construction of Global Knowledge by the Intergovernmental Panel on Climate Change. *International Environmental Agreements*, 5 (2: June), pp. 191–210.

________. (2004), "The Local, the Global, and the Kyoto Protocol" in S. Jasanoff and M.L. Martello, eds. *Earthly Politics, Worldly Knowledge: Local and Global in Environmental Politics*. Cambridge, MA: MIT Press, pp. 103–126.

________. (2002), *Greening The Earth With Trees: Science, Storylines And The Construction Of International Climate Change Institutions*. Doctoral Dissertation, Environmental Studies: University of California, Santa Cruz.

Foucault, M. (1980), *Power/Knowledge: Selected Interviews and Other Writings 1972–1977*, C. Gordon, ed. New York: Pantheon Books.

Gibbs, W.W. (1995), Lost Science in the Third World. *Scientific American*. (August), pp. 92–99.

Global Environmental Assessment Project (1997), *A Critical Evaluation of Global Environmental Assessments: The Climate Experience*. Calverton, MD: CARE.

Gupta, J. (1995), The Global Environmental Facility in Its North-South Context. *Environmental Politics*, 4 (1), pp. 19–43.

Haas, P. M. (2004), When does power listen to truth? A constructivist approach to the policy process. *Journal of European Public Policy*, 11 (4), pp. 569–592.

Herrick, C.N. and Jamieson, D. (2000), Junk Science and Environmental Policy: Obscuring Public Debate with Misleading Discourse. *Philosophy and Public Policy Quarterly*, (21: Spring), pp. 11–16.

Huntingford, C. and Gash, J. (2005), Climate Equity for All. *Science*, 309 (5742), pp. 1789.

IAI (Inter-American Institute) (2006), <http://en.wikipedia.org/wiki/Inter-American_Institute_for_Global_Change_Research> [accessed June 10].

Intergovernmental Panel on Climate Change (2000), *Special Report on Land Use, Land Use Change, and Forestry*. Cambridge: Cambridge University Press.

Jasanoff, S. (2004), "The Idiom of Co-Production" in S. Jasanoff, ed. *States of Knowledge: The Co-Production of Science and Social Order*. London: Taylor & Francis, Inc, pp. 1–12.

________. (1996), "Science and Norms in Global Environmental Regimes" in F.O. Hampson and J. Reppy, eds. *Earthly Goods: Environmental Change and Social Justice*. Ithaca and London: Cornell University Press, pp. 173–197.

________. (1990b), American Exceptionalism and the Political Acknowledgment of Risk. *Deadalus*, (Fall), pp. 61–81.

________. (1990a), *The Fifth Branch: Science Advisors as Policymakers*. Cambridge, MA: Harvard University Press.

Jasanoff, S. and Long Martello, M., eds. (2004), *Earthly Politics: Local and Global in Environmental Governance*. Cambridge, MA: MIT Press.

Jasanoff, S. and Wynne, B. (1998), "Science and Decisionmaking," in S. Rayner and E.L. Malone, eds. *Human Choice and Climate Change Volume One*. Columbus, Ohio: Batelle Press, pp. 1–87.

Johnson, K. (2001), Brazil and the Politics of the Climate Change Negotiations. *Journal of Environment and Development*, 10 (2), pp. 178–206.

Kandlikar, M. and Sagar, A. (1999), Climate Change Research and Analysis in India: An Integrated Assessment of a South-North Divide. *Global Environmental Change*, 9, pp. 119–138.

Lahsen, M. (forthcoming), "Knowledge, Democracy and Uneven Playing Fields: Insights from Climate Politics in – and between – the US and Brazil" in N. STEHR, ed. In *Knowledge and Democracy*. Transaction Publishers.

________. (2006), *Ceding Ground to Scientific Authority? Power and Perceptions of Science in the Climate Regime*. Paper presented at the International Studies Association, San Diego, California, 23 March.

________. (2005a), Technocracy, Democracy and US Climate Science Politics: The Need for Demarcations. *Science, Technology and Human Values*, 30(1), pp. 137–169.

________. (2005b), *Tattered or armed by science? Science and sovereignty in Brazil*. Discussion paper presented at the Science and Democracy Network Workshop, Harvard University, Cambridge, USA, 25 June.

________. (2004), "Transnational Locals: Brazilian Experiences of the Climate Regime" in S. Jasanoff and M.L. Matello, eds. *Earthly Politics: Local and Global in Environmental Politics*. Cambridge, MA: MIT Press, pp. 151–172.

________. (2001), *Brazilian Epistemers' Multiple Epistemes: An Exploration of Shared Meaning, Diverse Identities, and Geopolitics, in Global Change Science.* Belfer Center for Science and International Affairs (BCSIA) Discussion Paper 2002–16, Environment and Natural Resources Program <http://www.ksg.harvard.edu/gea/pubs/2002–01.htm>.

________. (1998), "The Detection and Attribution of Conspiracies: The Controversy Over Chapter 8" in G.E. Marcus, ed. *Paranoia Within Reason: A Casebook on Conspiracy as Explanation. Late Editions 6, Cultural Studies for the End of the Century*. Chicago: University of Chicago Press, pp. 111–126.

Lahsen, M. and Öberg, G. (2006), *The Role of Unstated Mistrust and Disparities in Scientific Capacity*. Report published by The Swedish Institute for Climate Science and Policy Research, Linköbing University, Sweden. <http://www.cspr.se/content/1/c6/05/43/37/CSPR%20Rapportserie.doc>.

Litfin, K. (2000), Environment, Wealth and Authority: Global Climate Change and Emerging Modes of Legitimation. *International Studies Review*, 2 (2), pp. 119–148.

Liverman, D. and O'Brien, K. (2002), "Southern Skies: The Perception and Management of Global Environmental Risks in Mexico" in Social Learning Group. *Learning to Manage Global Environmental Risks: Volume 1. A Comparative History of Social Responses to Climate Change, Ozone Depletion, and Acid Rain.* Cambridge, MA: MIT Press, 213–234.

Lutes, M. (2006), Vitae Civilis Institute, São Paulo, Brazil, personal conversation with Myanna Lahsen on August 21.

Marcus, G.E., ed. (1995), *Technoscientific Imaginaries: Conversations, Profiles, and Memoirs*. Chicago: Chicago University Press.

McNie, E. (forthcoming), Reconciling Supply and Demand of Scientific Information: A Review of the Literature. *Environmental Science and Policy*.

Miles, E.L., A. Underdal, S. Andresen, J. Wettestad, J.B. Skjærseth and Carlin, E.M. (2001), *Environmental Regime Effectiveness: Confronting Theory with Evidence*. Cambridge, MA: MIT Press.

Miller, C.A. (1998), *Extending Assessment Communities to Developing Countries*. ENRP Discussion Paper E-98-15. <http://www.ksg.harvard.edu/gea/pubs>.

Miller, C.A. and Edwards, P. N., eds. (2001), *Changing the Atmosphere: Expert Knowledge and Environmental Governance*. Cambridge, MA: MIT Press.

Ministry of Science and Technology (Brazil) (2006), Statistics obtained from <http://dgp. cnpq.br/censo2004/sumula_estat/index_pesquisadores.htm>.

Mitchell, R.B., W.C. Clark, D.W. Cash and Alcock, F. (2005), *Global Environmental Assessments: Information, Institutions, and Influence*. Cambridge, MA: MIT Press.

Mol, A.P. J. (2002), Ecological Modernization and the Global Economy. *Global Environmental Politics*, 2 (2), pp. 92–115.

Najam, A. (2005), "Why Environmental Politics Look Different from the South" in P. Dauvergne, ed. *Handbook of Global Environmental Politics*. Edward Elgar Pub, pp. 111–126.

Najam, A., S. Huq and Sokona, Y. (2003), Climate Negotiations Beyond Kyoto: Developing Countries Concerns and Interests. *Climate Policy*, 3 (3), pp. 221–231.

O'Riordan, T., C.L. Cooper and Jordan, A. (1998), "Institutional Frameworks for Political Action" in S. Rayner and E.L. Malone, eds. *Human Choice and Climate Change Volume One*. Columbus, Ohio: Batelle Press, pp. 345–439.

Oreskes, N. (2004), Science and Public Policy: What's Proof Got to Do with It? *Environmental Science and Policy*, 7, pp. 369–383.

Payne, R.A. (2001), Persuasion, Frames and Norm Selection. *European Journal of International Relations*, 7 (1), pp. 37–61.

Pielke Jr., R.A. (1994), Scientific Information and Global Change Policymaking. *Climatic Change*, 28, pp. 315–319.

Powell, W.W. and DiMaggio, P. J. (1991), *New Institutionalism in Organizational Analysis*. Chicago: University of Chicago Press.

Proctor, J.D. (1998), The Meaning of Global Environmental Change: Retheorizing Culture in Human Dimensions Research. *Global Environmental Change*, 8 (3), pp. 277–248.

Putnam, R.D. (1993), *Making Democracy Work: Civic Traditions in Modern Italy*. Princeton, NJ: Princeton University Press.

Rayner, S., D. Lack, H. Ingram and Houch, M. (2002), Weather Forecasts Are for Wimps: Why Water Resource Managers Don't Use Climate Forecasts. *NOAA Office of Global Programs*. <www.ogp. noaa.gov/mpe/csi/econhd/1998/rayner_final.pdf>.

Sarachick, E. S. and Shea, E. (1997), End-to-End Seasonal-to-Interannual Prediction. *The ENSO Signal*, 7, pp. 4–6.

Sarewitz, D. (2000), "Science and Environmental Policy: An Excess of Objectivity" in R. Frodeman, ed. *Earth Matters: The Earth Sciences, Philosophy, and the Claims of Community*, Prentice Hall, pp. 255–275. <http://www.cspo.org/products/articles/excess.objectivity.html>.

Shapin, S. (1995), Cordelia's Love: Credibility and the Social Studies of Science. *Perspectives on science*, 3, pp. 255–275.

Siebenhüner (2003), The Changing Role of Nation States in International Environmental Assessments – the Case of the IPCC. *Global Environmental Change*, 13 (2), pp. 113–123.

Smith, W. and Kelly, S. (2003), Science, Technical Expertise and the Human Environment. *Progress in Planning*, 60, pp. 321–394.

Steinberg, P. F. (2001), *Environmental leadership in developing countries: Transnational relations and biodiversity policy in Costa Rica and Bolivia*. Cambridge: MIT Press.

Stern, P. C. and Easterling, W. E. (1999), *Making Climate Forecasts Matter*. Washington, D.C.: National Academy Press.

Susskind, L.E. (1994), *Environmental Diplomacy: Negotiating More Effective Global Agreements*. New York, Oxford: Oxford: Oxford University Press.

Tesh, S. and Paes-Machado (2004), Sewers, Garbage, and Environmentalism in Brazil. *Journal of Environment and Development*, 13 (1), pp. 42–72.

VanDeever, S.D. (2005), "Effectiveness, Capacity Development and International Environmental Cooperation" in P. Dauvergne, ed. *Handbook of Global Environmental Politics*. Edward Elgar Pub, pp. 95–110.

Wennerås, C. and Wold, A. (1997), Nepotism and Sexism in Peer-Review. *Nature*, 22 (387: May), pp. 341–43.

Williams, M. (1993), Re-articulating the Third World coalition: The role of the environmental agenda. *Third World Quarterly*, 14 (1), pp. 7–29.

________. (2005), The Third World and Global Environmental Negotiations: Interests, Institutions and Ideas. *Global Environmental Politics*, 5 (3), pp. 48–69.

Yearley, S. (1996), *Sociology, Environmentalism, Globalization*. London & Thousand Oaks, CA: Sage.

Chapter 9

Disrupting the Global Discourse of Climate Change: The Case of Indigenous Voices

Heather A. Smith

Introduction

In May 2005, I attended a conference in Montreal, sponsored by Natural Resources Canada, on the topic of adapting to climate change. During that conference there were several sessions on climate impacts in the Arctic. One of the speakers was an Inuit hunter, John Keogak, from the Western Arctic (see Knotsch, Moquin and Keogak 2005). He gave one of the most compelling testimonies about climate change impacts that I have ever had the privilege to hear. He spoke of changing wind patterns that affected ice flows, of the change to his home and way of life, and his fear that there is no future for his grandchildren. He also challenged us, "the south and the wealthy" to remember the lives of those in Canada's north. This eloquent gentleman put a much needed human face on the issue of climate change. He reminded me of the voices, human and natural, that are too often marginalized from the world of targets, timetables, science and international negotiation. He challenged my/our Western[1] ways of knowing through his expression of traditional environmental knowledge (TEK), counseled us all to remember our connection to the land, and turned my/our sense of geographic space upside down as he spoke from the north to the south and from the local to the global.

In his talk, John Keogak sent messages with a clarity and eloquence that I cannot match. And yet, in some fashion, I seek to deliver similar messages.[2] While I am

1 Western ways of knowing is assumed here to include assumptions that privilege and value science over traditional indigenous knowledges. It is a construction of knowledge that separates us from the environment and also promotes a notion of economic progress that presumes domination over our resources. This divide is found as well in several chapters in this book including Bäckstrand and Lövbrand's discussion of ecological modernization as contrasted by William Smith's discussion of Totonac cosmology.

2 As a Western woman in Northern British Columbia I understand that this approach has ethical implications. While I have been adopted into the Bear Clan of the Carrier Sekani people, the indigenous voice is not my voice and I run the risk of engaging in a colonizing act through the creation of an "exotic other" (Levi and Dean 2003, 27). This is not my intention

neither an Inuit woman nor Canadian First Nations, I hope to draw on the words of indigenous peoples to support a broader case about the masking effect of the "global" social construction of climate change. The definition of climate change as global masks alternative voices that fundamentally challenge Western ways of knowing, being and doing. It will be seen that by highlighting indigenous voices, the climate change discourse becomes far more complicated as concepts embedded in the dominant global discourse, such as multilateralism, the state, and western ways of knowing, are problematized.

This chapter shows how indigenous voices provide a counterhegemonic discourse to the "global" Western construction of climate change. The starting point for the chapter is a discussion of the theoretical foundations that underpin the analysis. I eschew the traditional paradigms of realism and liberalism in favor of post-positivist critical and feminist theory. These theories are explicitly emancipatory and consequently fit well with the aim of highlighting a counterhegemonic discourse, such as that offered by indigenous peoples.

Next, I investigate the use of "global" as a concept used to justify particular kinds of actions and inaction, and which privileges a particular mode of thought and action (see also Paterson and Stripple, and William Smith this volume). Then, and similarly to Cathleen Fogel (2004), the chapter moves on to highlight the alternative climate change discourse as articulated by indigenous voices, with numerous references to the voices of northern[3] indigenous peoples.[4] The chapter concludes with the implications of the analysis and argues that to remain deaf to the counterhegemonic discourse is to deny the realities of climate change. Indigenous voices provide a genuine, if challenging, alternative.

Theoretical Foundations

Consistent with the work of preeminent critical theorist, Robert Cox, I assume that "there is … no such thing as theory in itself, divorced from a standpoint in time

whatsoever. The evidence here is all secondary sources and I present the voices of indigenous peoples as their own. Moreover, this work is respectfully written from the position of a political ally.

3 Northern in this instance represents a geographical space, peoples whose traditional lands are located in what is typically constructed as the "developed north" in contrast to the underdeveloped south. In some instances in this chapter, north/northern is invoked by indigenous peoples to indeed represent the developed northern states, in contrast to the less developed southern states.

4 The use of the terms indigenous, indigenous peoples', and First Nations, are not without consequence (see Keal, 2003). These are essentially contested and highly political identity constructs often historically associated with colonizing acts. However, indigenous peoples' has been adopted by a diverse set of peoples as a broad label from which to advocate for an alternative vision of the issue of climate change. As noted in Levi and Dean, "subaltern groups regularly reclaim ethnonyms applied to them by outsiders, often redefining old colonial labels with newly empowered meanings of self-affirmation" (2003, 8)

and space" (Cox and Sinclair 1996, 87; see also Smith 2004, 503). I reject problem solving theory which is understood to be theory which "takes the world as it finds it, with the prevailing social and power relationships and the institutions into which they are organized, as the given framework for action" (Cox 1986, 208). Critical theory, in contrast, "stands apart from the prevailing order of the world and asks how that order came about. Critical theory … does not take institutions and social and power relations for granted but calls them into question" (ibid.). Rather, we are encouraged to question institutions, social and power relations. We are reminded that "theory is always for someone and for some purpose" (Cox and Sinclair 1996, 87). Thus, from a critical perspective we are encouraged to ask for whom? For whom is the discourse? Who has constructed it and what interests does it serve to protect and promote? Who is excluded? By taking this approach we begin to challenge the dominant discourse, and we are able to begin to reveal what and who has been "silenced by the discourse."

Critical theory also "allows for a normative choice in favor of a social and political order different from the prevailing order, but it limits the range of choices to alternative orders which are feasible transformations of an existing world" (Cox and Sinclair 1996, 90; see also Linklater 1996). The theory does not cloak itself in the language of "legitimate social science" and "objectivity" as a means by which to avoid "normative or moral stances" (Smith 2004, 507). Thus, this chapter is informed by an explicit normative agenda, catalyzed by a "willingness to question all social and political boundaries and all systems of inclusion and exclusion" (Linklater 1996, 286).

These very broad themes are related to feminist theory through the work of Cynthia Enloe and Christine Sylvester. Both critical theory and feminist theory (or at least the scholars included here) are rooted in post-positivist epistemologies, both have embedded normative projects, and both reject the hegemonic vision of what constitutes appropriate theorizing in International Relations. The work of Enloe and Sylvester has particular merit because they interrogate the discipline and remind us of the everyday.

Christine Sylvester (1996, 262) has raised the issue of "everyday forms of feminist theorizing" which is understood to encompass "everyday forms of resistance and struggle [which] issue from activities of average people." She suggests that we look at people in their everyday places of action, places that would usually not attract the attention of the International Relations scholars. And she notes "to suggest bringing such people into international relations is earthshaking for a field that admits only official decision-makers, soldiers, statesmen, terrorists, kings, and the occasional 'crazed' religious group to the fold" (Sylvester 1996, 264).

The work of Cynthia Enloe affirms the value of this approach as she links the personal and the international and, in doing so, reveals the power of the traditional state-centric point of view, a point of view which dominates traditional theorizing in our field. The focus on "the state" limits our field of vision. Smith (2004, 504) argues that the state has been the focus of our analysis rather "than either humanity as a whole or the individual." Smith (2004, 505) continues, "International Relations theory has almost been defined by its worship of the state-as-actor, and the consequence downplaying of the role, or fate, of individuals or other actors." Enloe, in her assessments of the lives of women, reveals the fiction of "the state" and shows how much power is needed to keep people silent and marginalized (this brings an interesting perspective to the concept of territorialization introduced by the Stripple and Paterson chapter in this book).

Enloe further argues that, in the tradition of international relations, we have prized analytical economy and, in doing so, have underestimated the variety of power, "So omitting myriad strands of power amounts to exaggerating the simplicity of the entire political system. Today's conventional portrait of international politics thus too often ends up looking like a Superman comic strip, whereas it probably should resemble a Jackson Pollack" (Enloe 1996, 188–189). As a result of the way some regard the world of international relations, stuck in black boxes and levels of analysis, we remain immune from the Jackson Pollack nature of our world. "The margins stay marginal, the silent stay voiceless, and the ladders are never turned upside down" (Enloe 1996, 189).

In the case of climate change, it is the voices of the elites, states and sometimes regions that are typically represented in multilateral international negotiations, while indigenous peoples and their counterhegemonic discourse are masked and marginalized. The counterhegemonic discourse is dangerous to the status quo because it gives life to the theoretical claims noted above. As will be seen, the indigenous peoples' discourse questions institutions and social orders. Similar to Sylvester, the indigenous discourse shifts our attention to the everyday and like Enloe, their voices undermine statecentric assumptions. By questioning social order and looking at the everyday, we turn the ladders upside down and reveal the "Jackson Pollack" way of the world.

Climate Change is a Global Issue

Prior to an assessment of how the indigenous peoples' discourse disrupts the status quo, we need to consider the power of the construction of climate change as a "global" issue (see also Paterson and Stripple, and William Smith this volume). Climate change is routinely framed as a global environmental problem, which for some may appear unproblematic. However, the "climate change as global" perspective is not without implications. "Global problems, it is argued, require global solutions. This simple aphorism hides a multitude of political and economic difficulties, not the least of which is how we should understand the 'global' as an organizing principle" (Elliott 1998, 3). "Global" embodies a particular social order, complete with institutions, and ways of doing and knowing. Thus, following Elliott's observations, this section considers the power of construction of climate change as global. Consecutively this section considers these constructions as interlocking, but not necessarily consistent: global as a mechanism by which to universalize environmental threats; global as a means by which to externalize the environment, deny the local and provide a sense of distance and detachment; global as a solution as embodied in multilateral processes that are statecentric; the global other. It will be seen that there is power in the language of global to obfuscate power relations and prevailing social orders, to construct "others" and to deny responsibility.

Global Universalized Environmental Threats

First, the notion of "global" can be used to universalize threats. To suggest that climate change is a global issue implies that climate change is equally important to all parties. "The sense of the 'global' has been captured in metaphors intended to reflect our shared vulnerability to and responsibility for global insecurities—our common future, a global partnership, and our global neighbourhood" (Elliott 2002, 66). Yet, issues such as climate change are a reflection of interests of particular sets of states, who initially defined the problem as urgent, even if they have backed off on their commitments since the early political life of climate change. "The 'global' then is not simply a geographic term appropriated to describe increasing environmental independence ... rather it is a political term and one which ... provides the North with a new political space in which to control the South, thus creating the moral base for green imperialism" (Elliott 1998 citing Shiva, 252; see also Jasanoff 2004).

Globalized External Environment

Second, in much of the mainstream climate change discourse and practice, the notion of global reinforces a particular gaze which marginalizes and externalizes nature, making it subordinate. The marginalization of nature is linked to the global by Cathleen Fogel in her discussion of Karen Litfin's work (1997) on the "global gaze," described as "a fundamentally maculinist perception of the natural world as seen from a position of detachment and power" (Fogel 2004, 106). William Rees (2000, 21) observes that even the language of "the environment" and environmental studies, is often designed to keep us "separate and apart from nature." The environment is externalized while, in fact, according to Rees (2000, 23), if we seriously looked at our relationship with nature, we would see that we are indeed parasites.

The global is also used to obscure the local. By defining the issue as global, we are able to define the issue as something that is "out there" and not necessarily connected to our lives. However, our activities on a daily basis are part of that global problem, and we need to take responsibility for our actions. Yet we often engage in what Paul Wapner (2000) calls ecological displacement. The spatial dimension of ecological displacement occurs "when people are unwilling to address environmental problems as they emerge within their immediate community. As a result, people tend to export the experience of environmental harm to others" (Wapner 2000, 358). Spatial ecological displacement functions to deny local contributions to a problem and protect the interests of the status quo and, oftentimes the everyday consumer from the reality of the need to change patterns of consumption.

The marginalization of nature and the obscuring of the local are facilitated by denial and detachment which plays out in global climate processes. For example, symptomatically, there is a move to simplify nature through standardization and universalization with the end to "understanding" and "controlling" both natural and human objects (Fogel 2004, 106). If one considers international climate negotiations, one could argue that the negotiations are largely defined by regular denial about the minimal impact of Kyoto on GHG emissions and the often overbearing obsession with the economic impacts of emissions reductions strategies. As Fogel argues, the technical and the scientific are used to simplify nature and make it manageable; and in turn the climate change discourse becomes one about competitiveness while value is accorded to gross domestic product in an era when "more is better." It is "mechanistic, [and] market-oriented" (Fogel 2004, 121).

We can also see the detachment through how "first order questions" are crafted in international negotiations. Elliott (1998, 4) argues that there is a tendency to focus on targets, timetables, compliance, and transparency as though they are the first order questions. These questions, however, work often to shift our attention away from the broader social context wherein "the contemporary political and economic order is quite likely part of the problem" (Elliott 1998, 4).

Global and Multilateral Processes as Statecentric

Third, it is assumed that if there is a global problem, then there are often global solutions. In the case of climate change, the dominant feature has been negotiations through the UN Framework Convention on Climate Change (UNFCCC). Yet multilateralism cannot be treated unproblematically. From a critical perspective, we know that there are "many ways in which the prevailing order maintains its control of the debate by masking dangerous practices and packaging the debate in ways that obscure the destructive forces at work in the system" (Broadhead 2002, 20). Multilateralism, for example, can be seen a tool used to enforce hierarchy and define rules in the interests of particular states (see Cox 1997a, xvi). What is crucial, often embedded in the definition of climate change as global necessitating the search for global solution, is the assumption that these are state-based negotiations. This rather innocuous aspect of multilateralism is of great significance because it privileges a particular set of international actors (see also Stripple and Paterson this volume).

The Global Other

Finally, the global is used in particular ways to construct the "other." Who and what constitutes the other is fluid; however, in each case, there is a sense of threat related to the other. The Canadian government, for example, has stated that "no one country, acting alone, can solve the problem of climate change, but by working together towards a common goal, the nations of the world can successfully address the challenge" (Government of Canada 2002, 111). In making the case for global action the United States identified the need for developing countries to act and to engage in meaningful participation, because otherwise action on climate change is futile (see Elliott 1998, 65; Depledge 1999, 18). Thus, a global Kyoto meant all states participated, not just the historically heavy emitters. Again, referring to the Canadian case, the Government of Canada, notes that "it is anticipated that in future rounds, major developing countries will assume emissions targets, moving us closer to a truly global response" (Government of Canada 2002, 7). This is not to suggest that either Canada or the US are genuinely committed to accepting responsibility for their emissions history, but rather that the language of "global" or "truly global" is used to cast developing states in a light that implies they are not taking responsibility for a global issue. There is an attempt to cast developing states as environmental villains, when in actuality the developed states' emphasis on the global and the requirement for global action is a reflection of the economic interests which drive the action of many developed states (this theme is also discussed in the Lahsen, and Bäckstrand and Lövbrand chapters in this volume).

The embedded assumption that if Kyoto is not global, then developed states will lose economically can also be linked to the use of the term "global" and reveals a denial of responsibility. For example, when we focus upon the global in discussions of solutions to climate change, we find an emphasis on the need to reduce GHG concentrations, as indicated in the UNFCCC. It is well understood that the reduction of GHG concentrations and the emissions targets, as outlined in the Kyoto Protocol, will not fundamentally alter the warming trends. Furthermore, one of the reasons that the oft cited requirement of emissions reductions from 60–80 percent below 1990s levels will not even be considered is the lack of political will by key emitting states with historical responsibility for the present situation.

By casting the problem as global, there is a denial of historic responsibility for emissions; and by focusing on global concentrations, developed states are able to hide behind percentages that do not reflect their per capita emissions. This is the case with Canada that routinely argues that it is only responsible for 2 percent of the world's GHG emissions. The "climate change is a global issue" argument has also been used to bolster claims by Canada and other developed states that they should have the right to use the flexibility mechanisms in the Kyoto Protocol to take action outside of their states and get credit for those actions (as seen in the chapters by Cass, Pettenger and Hattori, the norm of economic efficiency supports the proposition that if we can reduce emissions elsewhere more cheaply then we should do so). Some have seen this as offloading responsibility to avoid domestic reductions (see H. Smith, 2004).

Finally, the global other has been constructed as international environmental governance. While global is used as a means by which to design rules in the interests of particular elites, the nature of the issue and global governance have been used to justify the rejection of the Kyoto Protocol. During the ratification debate in Canada in 2002, anti-Kyoto forces argued in favor of a made-in-Canada approach to climate change, claiming that the Kyoto Protocol was too onerous and did not reflect Canadian interests. The global construction (literally and figuratively) of the Kyoto Protocol was cast as anti-Canadian. Thus, the global is used to justify inaction, while simultaneously being used to blame "future polluters" for their inaction. This is the same kind of logic that informed American withdrawal from Kyoto.

The global has been used and manipulated to the ends of particular interests. This section has considered the dominant discourse and the construction of global as a means by which to universalize environmental threats, as well as the construction of the global as a means by which to externalize the environment, obscure the local, and facilitate detachment from the problem through Western scientific and economic ways of knowing. The link between global problems and solutions via multilateralism was also considered. Finally, the construction of the global other was examined, in particular the ways in which the concept global has been used to demonize developing states, support the denial of responsibility for historic emissions and, in some cases, to argue in favor of opting out of international environmental governance processes. In the next section, we turn to challenge the dominant discourse with the declarations of the Indigenous Peoples Caucus made between 1998 and 2004.

Indigenous Peoples and Climate Change

In terms of global environmental governance, there is some inclusion of indigenous peoples in several processes: the World Commission on Environment and Development, the United Nations Environment and Development as well as Agenda 21/Rio Declaration (Elliott 1998, 165), and the Johannesburg Summit Declaration. Indigenous peoples of the north also have permanent representation at the Arctic Council.

Indigenous people are in neither the UNFCCC nor the Kyoto Protocol. Yet, indigenous peoples have actively sought to have their voices heard in the climate change process. What follows are eight declarations made by indigenous people regarding climate change. The declarations are summarized at points, but I have included the original words as well so that their voices may be better heard.

First, the Albuquerque Declaration of 1998, produced by Native American groups, presented to the Fourth of Conference of the Parties, states that Indigenous peoples have known for a long time that the earth was in danger. They invoke traditional knowledge and articulate the following prediction:

> Our prophecies and teachings tell us that life on earth is in danger of coming to an end. We have accepted the responsibility designated by our prophecies to tell the world that we must live in peace and harmony and ensure balance with the rest of Creation. The destruction of the rest of Creation must not be allowed to continue, for if it does, Mother Earth will react in such a way that almost all people will suffer the end of life as we know it (Albuquerque Declaration, 1).

In addition, the Declaration is highly critical of the dominant economic and political orders. It includes claims to the right to self-determination, and the right and responsibility to control traditional knowledge. Canada and the US are criticized for lack of inclusion of indigenous peoples on their delegations for COP4. Lastly, there is a call for the "establishment and funding an Inter-sessional open ended working group for Indigenous Peoples within the Conference of the Parties" (ibid., 4), thus demanding equal partnerships at all levels of decision making.

Second, in May of 2000, representatives of indigenous organizations from around the world came together to produce the Quinto Declaration on Climate Change Negotiations (2000). The Quinto Declaration outlined a series of specific requests related to the participation of indigenous peoples in the UNFCCC. Quickly after the Quinto Declaration, the Declaration of the First International Forum of Indigenous Peoples on Climate Change (2000) was produced at the same time as the 13th Session of the Subsidiary Bodies to the UNFCCC was meeting in Lyon, France. The Forum statement is highly critical of Western development models:

> The scientists of Western society have dismissed us as sentimental and superstitious and accused us of being an obstacle to development. Paradoxically, those that previously turned deaf ears to our warnings, now are dismayed because their own model of "development" endangers our Mother Earth (2000, 1).

The forum statement demands respect for indigenous rights and identifies common positions on sinks in the clean development mechanism, public participation, land use, land use change and forestry, the adaptation fund, Joint Implementation, capacity building, and compliance. Concerning sinks in the Clean Development Mechanism (CDM), for example, it is stated that:

> Sinks in the CDM would constitute a worldwide strategy for expropriating our lands and territories and violating our fundamental rights that would culminate in a new form of colonialism. Sinks in the CDM would not help to reduce GHG emissions, rather it would provide industrialized countries with a ploy to avoid reducing their emissions at source (ibid., 2).

The statement represents the work of indigenous organizations from Peru, Panama, Nepal, Samoa, Rwanda, Colombia, Indonesia, Ecuador, Venezuela, and the Philippines.

Third, the Forum of Indigenous Peoples and Local Communities on Climate Change, representing the same group as noted above, presented a submission to this session of the subsidiary bodies (Forum of Indigenous Peoples and Local Communities on Climate Change 2000). This document is a fuller version of the principles noted above. The Kyoto Protocol is described as insufficient to avert climate catastrophe, and it clearly states that indigenous peoples are most vulnerable. This submission is also severely critical of Northern[5] life styles and consumption patterns.

---

5 As noted previously, Northern, in this instance, represents the wording of the Indigenous Peoples Caucus and is meant to represent the industrialized North which would typically include Canada, the United States, Europe.

> We believe that those that harm the Earth should bear responsibility for healing it, regardless of the cost. Over-production and over-consumption in the North have critically damaged Earth's life-support systems resulting in the current planetary crisis (ibid., Article 1.5).

Annex I states are criticized for the CDM as a means by which to avoid domestic emissions reductions (Article 3.2) and sinks provisions are rejected as having an "immoral and perverse dimension that, according to the indigenous cosmovision, do not stop the causes of climate change nor do they oblige the polluters to assume all the responsibilities" (ibid., Article 4.3). In addition, the sacred is invoked: "For Indigenous Peoples and local communities, the land is a sacred space that the Creator entrusts to our Peoples, hence our responsibility to care for it" (ibid., Article 1.12).

Fourth, the declaration of the Second International Indigenous Forum on Climate Change of November 2000 at COP6 at the Hague articulates views consistent with those above. Indigenous spirituality is presented as a primary consideration. It is stated:

> Earth is our Mother. Our special relationship with Earth as stewards, as holders of indigenous knowledge cannot be set aside. Our special relation with her has allowed us to develop for millennia a particular knowledge of the environment that is the foundation of our lifestyles, institutions, spirituality and world view. Therefore, in our philosophies, the Earth is not a commodity, but a sacred space that the Creator has entrusted us to care for her, this home where all beings live (The Hague Declaration of the Second International Forum of Indigenous Peoples and Local Communities on Climate Change 2000, 1).

The statement delivered to COP6 also rejects the inclusion of sinks within the CDM, calls for a Division on Indigenous Peoples within the Convention's Secretariat, proposes a moratorium on all oil, gas, nuclear and hydro-electric activities in pristine areas and blames industrialized states for climate change. This declaration includes signatories representing indigenous peoples around the world and beyond those noted above, including indigenous peoples from Russia, Mexico, Kenya, New Zealand, the United States, Canada and Guatemala.

Fifth, the Bonn Declaration of the Third International Forum of Indigenous Peoples and Local Communities on Climate Change (2001) was presented at the second session of the Sixth Conference of Parties. Again, one finds severe criticism of Western development models and a call for the recognition of indigenous rights to self determination and the right to full participation in the discussions of the UNFCCC. The Bonn Declaration criticizes neo-liberal valuation of nature:

> We openly oppose the measures to mitigate climate change under discussion that are based essentially on a mercantilist and utilitarian vision of the forests, seas, territories and resources of Indigenous Peoples, which are being valued for their capacity to absorb CO_2 and produce oxygen, and which negate our traditional cultural practices and spiritual values (ibid., 2).

The Declaration also includes a very strong statement about the inherent value of indigenous ways as well as concerns about racism:

> We are convinced that our philosophies and traditional practices are the most appropriate for the management of the ecosystems of our territories. Finally, we are also particularly concerned about the emergence of 'biocolonialism' and 'environmental racism' that Indigenous Peoples and Local Communities of the world continually confront (ibid., 3).

The Declaration concludes, "The damage caused by climate change exacerbates existing concerns and inequities and constitutes a matter of environmental and climate justice. The issue confronting humanity today is one of justice" (ibid.). Similar to the other declarations, the Bonn Declaration is signed by representatives of indigenous organizations from around the world including Columbia, New Zealand, the US, Kenya, Nepal, Thailand and Panama.

Sixth, the Indigenous Peoples and Local Communities Caucus (2001) statement to the Seventh Conference of Parties in Marrakech demanded the recognition of indigenous rights and as above, charged that the inclusion of sinks in the CDM was a mechanism for "a new form of colonialism." The Indigenous Peoples' Caucus Statement on Climate Change, COP8 (2002) also called for the recognition of indigenous rights and highlighted the impacts of climate change on indigenous peoples. The fact that the Kyoto Protocol will not stop climate change is not hidden or masked; rather, we are told: "The Kyoto Protocol is not sufficient to reverse, mitigate or stop the catastrophes that threaten our Mother Planet Earth" (2002, 1). This statement, like the others also calls for a Working Group of Indigenous Peoples on Climate Change, particular regulations related to the rights of indigenous peoples in the areas of the CDM and joint implementation, supports the adaptation fund, and calls for indigenous participation in UNFCCC capacity building initiatives.

Seventh, the Milan Declaration of the Sixth International Indigenous Peoples Forum on Climate Change (2003) reiterates the threat of climate change to their cultural and spiritual survival, "Our ancestral territories, spiritual, social, biological and cultural resources are the fundamental basis for our existence, health and livelihoods but are threatened and destroyed by climate change and its consequences." In the conclusion it is claimed that "we are securing not only our future, but the future of humanity and social and environmental justice for all" (ibid., 2).

Finally, the Declaration of the Indigenous Peoples Attending COP10, UNFCCC explicitly speaks on behalf of the global community of indigenous peoples, noting that the statement comes from "Indigenous peoples, globally, North and South" (2004, 1). The 2004 Declaration reiterates the call for the creation of a working group on Indigenous Peoples and calls for full participation of Indigenous Peoples. They ask why the working group has not been created and challenges states with the following question: "are we not part of this planet" (2004, 2)? Criticism is also leveled specifically at states: "As observers to the COP10, we are hearing from the states the same old arguments being discussed on how to alleviate and mitigate the climate disasters that affect all humanity" (ibid., 2). The Declaration also vigorously restates the vulnerability of Indigenous Peoples, arguing that "We are in crisis. We are in an accelerating spiral of climate change increasingly getting worse due to the conditions that industry, multinationals and governments impose upon our communities and the world" (ibid., 2). And finally, Mother Earth is included: "We consider this planet our Mother Earth where all humanity is born and nurtured. It is time that we looked to each other and that we listen to each other, recognizing and valuing the cultural and human qualities within each of us" (ibid.).

Challenging the Dominant Discourse

The statements of the indigenous peoples articulate a common set of themes that serve to disrupt the dominant way of thinking about climate change. It is now time to return to the set constructions of the global as articulated in the previous section of the chapter to explicitly compare them with the indigenous peoples' discourse, and to highlight the ways in which the indigenous voices provide for an alternative vision of climate change. We begin with the universalization of environmental threats.

Revisiting Universalized Environmental Threats

In almost every indigenous declaration noted above, the notion of climate change as a global problem of equal importance to all is challenged. Without question, the indigenous peoples' statements concur that climate change is a serious problem. They clearly articulate that we are in crisis. But the crisis they identify is one embedded in a completely dysfunctional economic order. The threat is not some abstract, long term environmental change. The threat is in the commodification of nature, and the production and consumption patterns of industrialized states. Moreover, indigenous peoples argue that they and their ways of life are especially vulnerable, citing for example, the Arctic and Arid regions. Recall the comments in the Indigenous Peoples' declarations about the CDM. Global claims are made in such a way that there is a perpetuation of a social and economic order that is racist and unsustainable. The political nature of the term, identified earlier, is exemplified in the various indigenous caucus declarations that point to bio-colonialism. The inequality is masked by arguing that climate change is global. There is no "common future." It is not global and equal; it is global and unequal.

Global as a Means of Domination and Suppression

One of the fundamental elements of the indigenous peoples' discourse is that it does not externalize the environment. It emphasizes the local (rather than denying it) and functions with a sense of attachment (rather than detachment) where nature is concerned. The global gaze is flipped as the indigenous vision is not one of dominance *over* nature, but rather a spiritual and cultural connection *to* nature. It positions Mother Earth to be respected and revered, not dominated and controlled.

Likewise, the local is revealed (not obscured). The testimonial noted at the beginning of the chapter emphasizes the weaving of local and the global together, and constantly reminds us to recognize the everyday and the human face, even while indigenous organizations are working at international negotiations. When indigenous peoples speak of the destruction of their lands and their cultures, they are speaking of their everyday and their local. The human and local are, for example, central to the observations of Sheila Watt-Cloutier, Chair of the Inuit Circumpolar Conference, of the negotiations at COP 9 in Milan:

> Global climate change negotiations are highly detailed and technical. There were thousands of delegates and I wondered how we could possibly inject the human dimension, the Arctic voice into the global debate. Had we forgotten the dramatic predictions of changes to our homeland, the Arctic; had we forgotten the Small Island States in the Pacific that may be underwater within my lifetime, had we forgotten the effects on our prairie farmers? The human impacts of climate change seemed to be lost in the technical detail (Inuit Circumpolar Conference 2004, 6).

The technical detail observed by Watt-Cloutier is symptomatic of the denial and detachment of the first order question of targets, timetables and compliance. The technical detail is also about a particular way of knowing, a Western way of knowing grounded in Western-based science. In contrast, the indigenous peoples' declarations continually make reference to traditional environmental knowledge which they have developed over the millennia and which helps them connect to Mother Earth. Similarly, Western development strategies are consistently questioned. The ways of being with nature and the ways of working with nature articulated in the indigenous peoples' declarations are opposed to those which are embedded in the dominant discourse. The indigenous ways problematize the Western denial and detachment of the global gaze.

Global as a Solution

The challenge of the indigenous peoples' caucus is to reclaim the global for their own. Recall the use of the notion of a global indigenous community in the Declaration of COP10. In that case the global is not statecentric, but a global defined by indigenous communities from the North and the South. An important implication of the definition of the global indigenous community is that "North" and "South" are problematized. For the Inuit and the Dene, Toronto, Dallas and Geneva are "the South." In locality they are southern, but conceptually they are considered part of the "northern" source of the problem. This counters the common tendency to equate indigenous peoples with the South.[6]

In addition, the various groups working within the indigenous caucus also force us to consider the importance of transnational ties in the definition of community. For example, a key player in the articulation of Northern indigenous views is the Inuit Circumpolar Conference (ICC). Founded in 1977 the ICC represents "approximately 150,000 Inuit of Alaska, Canada, Greenland and Chukotka (Russia)"

6 This observation is consistent with one made by Lorraine Elliot 1998, 158.

(Inuit Circumpolar Conference 2006, 1). The ICC is very outspoken and purposeful on climate change.

Finally, with reference to the statecentric bias of multilateralism, the indigenous discourse challenges the rightful voices in the multilateral fora. They challenge the states in which they are located; and through their demands to be heard at the UNFCCC, they remind us of their and our humanity. They have shifted the focus of global, from one dominated by the north, by simply asking "are we not part of this planet?" They move the dialogue from state-to-state to within the context of common humanity.

From this perspective, and building on the earlier discussion of multilateralism, the dominant order could mask dangerous practices. The declarations of the indigenous peoples, similar to the statements by many international non-governmental environmental organizations, reveal those dangerous practices. They remind us that Kyoto will not be sufficient to meet the challenge of climate change. They call carbon sinks immoral and perverse because they do not fundamentally alter Northern consumption patterns, and are regarded as a resource grab. Through their statements, they challenge not only the appropriate players in the multilateral game, they raise questions about the unsustainable practices designed through those negotiations and call our attention to the interests of the polluting elites who fashion those instruments.

The Global Other

The defiance embedded in the declarations of the indigenous peoples can be seen as a response to being silenced by the dominant discourse. There is a sense that indigenous peoples are the most vulnerable, but they are not willing to accept this position meekly. They will not allow environmental racism to go unchallenged. For example, Sheila Watt-Cloutier (2003), Chair of the ICC, attended COP9 in Milan and spoke to the press about the Inuit concerns. She argued that the human rights of the Inuit were under threat and presented evidence of the climate change in the North. She counseled against being dismissive of the Inuit, noting that "the Arctic is the barometer of the globe's environmental health. You can take the pulse of the world in the Arctic. Inuit, the people who live further north then anyone else, are the canary in the global coal mine."

Peoples, who are remote and marginalized, may become the canary in the global coal mine, but they are clear as to who constitutes their "other." Over and over again in the declarations above, the point is made that the villains in the case of climate change are developed states and that multinationals continue to act in ways that are unsustainable. More significantly, it is the production and consumption patterns of the wealthy that have put the world in crisis. Those of us in developed states are taken to task and called to take responsibility for our historic contribution to GHG emissions. Construction of the alternative indigenous discourse, upending the ladders per Enloe, is not about economic well being, leakage or competitiveness, and it situates the developed states as the other. The alternative global other is constructed on the basis of those who use the most and do the most harm to nature.

Concluding Reflections

This chapter began with the premise that the dominant climate change discourse focuses on global problems and global solutions, operates to privilege state-based interests and marginalizes indigenous voices. The environmental realities of climate change remind us that such privilege continues and has detrimental impacts. We are called by Robert Cox to consider that "Dominant forms of knowledge are nourished by dominant power. Dominant power orients the development of knowledge as an instrument for maintaining dominance, whether in the organization of production or in the control of deviance, whether in reinforcing the effectiveness of the state or of private capital" (Cox 1997b, 245).

There is much at stake if we genuinely challenge the global and pay heed to the themes that resonate throughout the declarations of the indigenous peoples. Some may be dismissive of the claims made by the indigenous peoples. Notions of Mother Earth and the Creator are hard to square with climate change targets and timetables. Yet, the world of targets and timetables, of CDM and JI, has not been particularly effective in terms of dealing with the problem of climate change; and perhaps the problem is precisely the dominant social order of which climate change negotiations are part. It would be naïve to suggest a dismantling of the social and economic order as we know it; and moreover, critical theory encourages us to consider feasible alternatives. The power of the alternative voice of the indigenous peoples is that, regardless of our entry into the post-Kyoto era, their concerns and the inequities they face continue to resonate. As such, and similar to the radical Green Governmentality discourse discussed in Bäckstrand and Lövbrand (this volume), they provide a starting point to reflect on our connections to the world as both individuals and scholars.

As scholars we must consider how we may reinforce the dominant discourse through our scholarly practices. While engaging in the research for this chapter it became clear that we too can become forces of homogenization. There are unfortunately too many examples of silencing of indigenous peoples in Northern developed states by scholars. Fogel (2004), for example, while producing an outstanding piece, does not even give a nod to indigenous peoples in the so-called North. Discussions of inequality have often focused on the geographical South while ignoring the complexities of the North. While small island states will suffer, so too will the peoples of the Arctic. The North and developed states too often become homogenous, denied is the place of the First Nations of Canada, the Inuit and the Dene. Far too often, there is an absence of mind regarding indigenous peoples in developed states and a simultaneous reinforcement of the statecentric billiard ball frame of the world (much like William Smith's "absence of mind" in this volume). The voices of indigenous peoples can also provoke us as individuals to consider our part in the perpetuation of environmental racism. To even evoke the concept of racism is to raise the hackles of many. But the question remains: how do our daily activities contribute to climate change and how does that have an impact on the lives of others?

Beyond our contribution to the problem, we should be encouraged to consider our own local. Through their connection to nature, many indigenous peoples are witnessing change in their environments. What are we witnessing? Do you see climate change? In Northern British Columbia we see climate change everyday as the mountain pine beetles are producing an epidemic that is devastating the forestry industry. Huge stands of pine trees are turning red before our eyes. The persistence of the pine beetle has been linked to warming trends. The "local" in Northern British Columbia is changing radically and yet many still remain detached from any sense of contributing to that change.

Indigenous peoples also show us that there are alternative ways of knowing about the environment. Traditional environmental knowledge, defined as "a cumulative body of knowledge, practice and belief, evolving by adaptive processes and handed down through generation by cultural transmission" (Berkes 1998, 8 as cited in Nichols et al 2004, 69), can help us understand more clearly the impacts of climate change. While there are questions around the relationship between scientific knowledge and TEK (Nichols et al, 2004), considerable work is being done related to TEK, climate change and Canada's Arctic. For example, the International Institute for Sustainable Development and the community of Sachs Harbor, Banks Island, Northwest Territories instigated a project that included community observations of climate change and produced a video which was taken to COP6 in the Hague (Fenge 2001, 5). Inuvialuilt in Sachs Harbor reported the following: "melting permafrost resulting in beach slumping; increasing snowfalls; longer sea ice-free season; new species of birds and fish ... near the community; a decline in the lemming population, the basic food for Arctic fox, a valuable harvested species; and generally a warming trend" (ibid.; see also Nichols et al 2004; Arctic Council 2004).

What if we pay heed to indigenous peoples and their concerns? We are then forced to ask (as also discussed in the William Smith chapter): how do we manage the sacred? In a world in which nature has become a commodity, we are told that we cannot commodify nature. The claims of ownership and rights to self determination are anathema to states which are developed and developing. Even the reference "indigenous peoples" is political as states such as Canada and Brazil lobbied to ensure the reference was "indigenous people" in Agenda 21, because they did not condone any implication of a community with rights (Elliott 1998, 166). The indigenous peoples' counterhegemonic discourse is itself socially constructed and reflects the political, economic and even spiritual claims made by indigenous peoples.

If we listened to indigenous peoples, we may have to acknowledge there is a crisis and that climate change is a life and death issue with profound ethical questions:

> Global warming ... raises the question directly whether some people will be allowed to gain at other people's expense, including their very life. Global warming will force us to

think more deeply about human duties to plants and animals than any other environmental crises, because the decision on atmospheric greenhouse gas stabilization levels will make us decide which plants and animals will survive (Brown 2003, 233).

Ultimately, indigenous voices remind us of our place and our humanity, and how humanity is connected within itself and to the earth. It is an uncomfortable reminder.

References

Albuquerque Declaration (1998), *Circles of Wisdom*. Native Peoples/Native Homelands Climate Change Workshop-Summit, Albuquerque, New Mexico. <http://www.ienearth.org/globalcc.html>.

Arctic Council (2004), *Impacts of Warming Arctic: Arctic Climate Impact Assessment*. Cambridge: Cambridge University Press.

The Bonn Declaration of the Third International Forum of Indigenous Peoples and Local Communities on Climate Change (2001), July 14–15. <http://www.tebtebba.org/tebtebba_files/susdev/cc_energy/bonndeclaration.html>.

Broadhead, L.A. (2002), *International Environmental Politics: The Limits of Green Diplomacy*. Boulder, Co: Lynne Rienner.

Brown, D.A. (2003), The Importance of Expressly Examining Global Warming Policy Issues Through an Ethnical Prism. *Global Environmental Change*, 13, pp. 229–234.

Cox, R. (1986), "Social Forces, States and World Orders: Beyond International Relations Theory" in R.O. Keohane, ed. *Neorealism and Its Critics*. New York: Columbia University Press, pp. 204–254.

________. (1997a), "Introduction" in R. Cox, ed. *New Realism: Perspectives on Multilateralism and World Order*. New York: United Nations Press, pp. 1–2.

________. (1997b), "Some Reflections on the Oslo Symposium" in S. Gill, ed. *Globalization, Democratization and Multilateralism*. New York: St. Martin's Press, pp. 245–252.

Cox, R and Sinclair, T. (1996), *Approaches to World Order*. Cambridge: Cambridge University Press.

Declaration of the First International Forum of Indigenous Peoples on Climate Change (2000), Lyon, France, September 4–6. <http://www.tebtebba.org/tebtebba_files/susdev/cc_energy/lyondeclaration.html>.

Declaration of the Indigenous Peoples Attending COP 10 (2004), UNFCCC. <http://www.tebtebba.org/tebtebba_files/susdev/cc_energy/buenosaires.html>.

Depledge, J. (1999), Coming of Age at Buenos Aires. *Environment*,41 (7), pp. 15–20.

Elliott, L. (2002), "Global Environmental Governance" in R. Wilkinson and S. Hughes, eds. *Global Governance: Critical Perspectives*. London: Routledge, pp. 57–74.

__________. (1998), *The Global Politics of the Environment*. New York University Press: New York.

Enloe, C. (1996), "Margins, Silences and Bottom Rungs: How to Overcome the Underestimation of Power in the Study of International Relations" in S. Smith, K. Booth and M. Zalewski, eds. *International Theory: Positivism and Beyond*. Cambridge: Cambridge University Press, pp. 186–202.

Fenge, T. (2001), The Inuit and Climate Change. *Isuma*, 2 (4: Winter). <http://www.isuma.net/v02n04/fenge/fenge_.shtml>.

Fogel, C. (2004), "The Local, the Global and the Kyoto Protocol" in S. Jasanoff and M.L. Martello, eds. *Earthly Politics: Local and Global In Environmental Governance*. Cambridge, Mass: MIT Press, pp. 103–126.

Forum of Indigenous Peoples and Local Communities on Climate Change (2000), *Position Paper Presented to the 13th Session of the Subsidiary Bodies to the UNFCCC, Lyon, France, September 4–15, 2000*.

Government of Canada (2002), *Climate Change Plan for Canada*. November. <http://www.climatechange.gc.ca/plan_for_canada/>.

The Hague Declaration of the Second International Forum of Indigenous Peoples and Local Communities on Climate Change (2000), November. <http://www.tebtebba.org/tebtebba_files/susdev/cc_energy/hague.html>.

Harris, P.G. (2002), Global Warming in Asia-Pacific: Environmental Change vs. International Justice. *Asia-Pacific Review*, 9 (2), pp. 130–149.

Indigenous Caucus of the Framework Convention on Climate Change (2003), *Indigenous Intervention on CDM.*, December. <http://www.indigene.de/cdm-intervention.html>.

The Indigenous Peoples and Local Communities Caucus (2001), *Seventh Session of the Conference of the Parties, United Nations Framework Convention on Climate Change*, Marrakech, 29 October – 9 November. <http://www.tebtebba.org/tebtebba_files/susdev/cc_energy/marrakech.html>.

The Indigenous Peoples' Caucus Statement on Climate Change, COP8 (2002), New Delhi, 23 October– 1 November. <http://www.tebtebba.org/tebtebba_files/susdev/cc_energy/hague.html>.

International Indigenous Peoples Forum On Climate Change (2003a), *Press Statement*, December. < http://www.treatycouncil.org/>.

________. (2003b), *Statement for Contact Group on Sinks*, December.

Inuit Circumpolar Conference (2004), Bringing Inuit and Arctic Perspectives to the Global Stage: Lessons and Opportunities. *Silarjualiriniq*, 19 (August to December). <http://www.inuitcircumpolar.com/index.php?ID=277&Lang=En>.

________ (2006), *ICC Overview*. <http://inuitcircumpolar.com/index.php?ID=16&Lang=En&Parent_ID=16>.

Jasanoff, S. (2004), "Heaven and Earth: The Politics of Environmental Images" in S. Jasanoff and M.L. Martello, eds. *Earthly Politics: Local and Global In Environmental Governance*. Cambridge, Mass: MIT Press, pp. 31–54.

Keal, P. (2003), *European Conquest and the Rights of Indigenous Peoples: The Moral Backwardness of International Society*. Cambridge: Cambridge University Press.

Knotsch, C., H. Moquin and Keogak, J. (2005), *Inuit Concerns of Changing Aquatic Ecosystems in Arctic Communities* and *Inuit Tapiriit Kanatami*. Conference on Adapting to Climate Change in Canada 2005: Understanding Risks and Building Capacity. <http://adaptation2005.ca/home_e.html>.

Levi, J.M. and Dean, B. (2003), "Introduction" in B. Dean and J.M. Levi, eds. *At the Risk of Being Heard: Identity, Indigenous Rights and Postcolonial States*. Ann Arbor: University of Michigan Press, pp. 1–44.

Linklater, A. (1996), "The Achievements of Critical Theory" in S. Smith, K. Booth and M. Zalewski, eds. *International Theory: Positivism and Beyond*. Cambridge: Cambridge University Press, pp. 279–300.

Litfin, K. (1997) The Gendered Eye in the Sky: A Feminist Perspective on Earth Observation Satellites, *Frontiers*, 18 (2), pp. 26–47.

Milan Declaration of the Sixth International Indigenous Peoples Forum on Climate Change (2003), 29–30 November. <http://www.tebtebba.org/tebtebba_files/susdev/cc_energy/milan.html>.

Nichols, T., F. Berkes, D. Jolly, N.B. Snow and the community of Sachs Harbour (2004), Climate Change and Sea Ice: Local Observations From the Canadian Western Arctic. *Arctic*, 57 (1), pp. 68–79.

Quinto Declaration (2000), Recommendations of Indigenous Peoples and Organizations Regarding the Process of the Framework Convention on Climate Change. <http://www.tebtebba.org/tebtebba_files/susdev/cc_energy/quito.html>.

Rees, W.E. (2000), "Ecological Footprints and the Pathology of Consumption" in R.F. Wollard and A.S. Ostry, eds. *Fatal Consumption: Rethinking Sustainable Development*. Vancouver: UBC Press, pp. 21–51.

Roberts, J.T. (2001), Global Inequality and Climate Change. *Society and Natural Resources*, 14.

Smith, H. (2004), "Seeking the Middle Ground Between More and Less: A Canadian Perspective" in V.I. Grover, ed. *Climate Change: Five Years after Kyoto*. Enfield, NH: Science Publishers, Inc., pp. 273–296.

Smith, S. (2004), Singing Our World into Existence: International Relations Theory and September 11. *International Studies Quarterly*. 48, pp. 499–515.

Sylvester, C. (1996), "The Contributions of Feminist Theory to International Relations" in S. Smith, K. Booth and M. Zalewski, eds. *International Theory: Positivism and Beyond*. Cambridge: Cambridge University Press, pp. 254–278.

Wapner, P. (2000), People, Nature, and Ethics. *Current History*, (November), pp. 355–360.

Watt Cloutier, S. (2003), Speech Notes for the Conference of Parties of the United Nations Framework Convention on Climate Change, 10 December. *Silarjualiring*, 17 (December), <http://www.inuitcircumpolar.com/index.php?ID=253&Lang=En>.

Chapter 10

Presence of Mind as Working Climate Change Knowledge: A Totonac Cosmopolitics

William D. Smith

Introduction: Presence of Mind in the Cosmological Weave

This chapter will depart from most of the others in the volume to examine a standpoint on climate change not from the Western scientists, politicians, activists, and media actors who dominate the global forum on the future of the planet, but from a non-Western people with a distinct land ethic, conception of community, and interpretation of climate issues. The analysis that follows, of climate change discourse in one indigenous smallholder farming community in east central Mexico, resonates with Heather Smith's call (this volume) to attend to "indigenous voices" in renewing approaches to climate change questions. This detailed case study focuses on a community far away from high profile climate change forums.

My methodological and ethical premise is that we need to expand the space of the "critical" in social constructionist approaches to knowledge, above all within the critical study of science, by bringing to "our" discussion the marginalized peoples standing right now on the front lines of climatic insecurity. The still-expanding interdisciplinary field of science and technology studies has leveled its gaze overwhelmingly on Western science; in contrast, we are just beginning to entertain other scientific traditions as critical interventions in a more inclusive discussion. I hope to show here that exploring understandings of climate change cross-culturally allows us to think in fresh ways about relationships between humans and the rest of nature, about what climate itself is and where it comes from, about what kinds of actors play key roles in weather patterns and events, and perhaps above all, about the nature of responsibility. In the debate on climate change, we (still) need to balance climate as a "statistical abstraction" (Demeritt 2001, 312) with a longer and harder look at on-the-ground, in-place human accountability (see Heather Smith this volume). More germane to the theoretical agenda of this volume, if social constructionist frameworks seek to understand how we are led to think about matters and with what real-world repercussions, then one especially provocative feature in the approach I will represent concerns the social and environmental stakes of "good" versus "bad" thinking.

The climate change discourse I examine belongs to a certain mode of thinking, one might better call it a cognitive faculty, among Totonac *campesinos* in the municipality of Huehuetla, an Indian-majority community in the Sierra Norte de Puebla, with a population of roughly 16,000. I should note from the outset that if one were to ask a Huehueteco farmer what his viewpoint might be regarding "global warming," he would likely respond with a blank stare. This does not mean he would have nothing to say about climate change; the Totonac farmers I know talk about that all the time. But in places like Huehuetla, climatic trajectories are imbricated in a local weave of the world, a local cosmology. One must approach local understandings of climate change, therefore, through the strands of the weave, some of which might strike "us" as distant indeed from the weather. For Huehuetecos "environment" (including climate) and "cosmos" mean much the same thing. That is, the environment reaches from the subsoil to the heavens and is animated by divine as well as human powers. The essential human role in the cosmos has much to do with qualities of thought. The Huehuetecos I present below speak a good deal about the necessity of "good thinking," a complex habit of mind with two essential dimensions. First, a sound thinker abidingly traces and attends to the connections among a wide array of cosmological domains, from soil to rain to drought to villages to gods. That is, Totonacs do not exercise knowledge to parse out the components of "nature," they exercise a kind of cognitive energy to see and thereby actually maintain the weave of the world. Second, good thinking is a practical consciousness with clear prescriptions regarding social relations. The field of the social is broad, embracing nearly all domains, that is to say from soil to gods. Thus good thinkers literally sustain the world of human and non-human as they hold in mind the relationships among those domains. Paraphrasing Totonac descriptions themselves of this habit of mind, I have termed this habit of good thinking as "presence of mind," which keeps this social world together in constructive relationships.

Although a Western view would define presence of mind as a religion-leavened notion of human faculty, Huehuetecos insist that it is an environmental science. The Totonac scientist, although a keen empirical observer, is not the hypothesis formulator and theory builder of the Western scientific paradigm. Rather, he or she possesses a feel for connections among people, environmental elements, and the will of the saints. She or he attunes thinking to the domain of soil, plants, water, and divinity. And he or she labors on the land accordingly. This is central: in analyzing presence of mind as pivotal to Totonac environmental science, I will place special emphasis on the character of labor, that is, the relationships between basic economic activity performed by the human body, good and bad thinking, and the environmental implications of those. It would be disingenuous to suggest that Totonac people in general, most of them classified as "extremely marginal" by government demographers and statisticians, pick and choose among the work options available to them. Totonac scientists do, however, distinguish good labor from bad. The former is frequently called righteous labor, inseparable from presence of mind.

Intersubjectivity and Actor Network Theory

Totonacs are by no means the only indigenous people to accord subjectivity to nonhumans. For many groups, plants, divinity, soil, water, and climate, among other elements, actively participate in conforming a world regarded essentially as social, one held together by mutual influence, reciprocity, channels of communication, and rights and obligations (Allen 1988; Kuletz 1998; Descola 1996). Here there is a degree of rapprochement between indigenous "folk theory" and one current in recent Western theory about the way the world is put together. In particular, nonhuman agents in Totonac cosmology might be classified as nonhuman "actants" by actor network theory (ANT) (Latour 2005, 1999, 1996, 1993; Law and Hassard 1999). ANT is one of the more influential network theories available to analysts who argue that orders, such as scientific discoveries or technological advances or whole socioecologies, must be understood as outcomes of human and nonhuman contributions. Thus network theories, from Latour to Deleuze and Guattari (1987) to Serres (1995, 1991, 1982) to Whatmore (2001), are a useful starting point within the constructionist literature for theorizing cosmologies like the Huehueteco. They are useful, too, for those of us interested in extending the reach of constructionist approaches to social and "natural" phenomena. I argue that social constructionism does not pay sufficient attention to nonhuman influence in order building, social and otherwise (Murdoch 2001). Climate change discourses are would-be orders. Scientists, lobbyists, and news networks construct them, but, in an apparently very real way, so does carbon dioxide.

Yet even ANT falls short when it comes to grasping nonhuman agency in its plenitude. As a framework geared largely to the modern and postmodern West, it has granted little space in its "heterogeneous field[s] of force" (Lee and Brown 1994, 782) to actors like patron saints and Madre Tierra. Also, in its insistence on placing human and nonhuman on equal footing, it is reluctant to entertain any privileged position for human agency. Hence the Totonac scientist would find ANT unequal to the task of comprehending presence of mind. It, as well as absence of mind, is a human attribute with a decisive influence on the fate of the cosmos. The knowledge of Huehuetla scientists comes close to what philosopher Gilles Deleuze (1993) describes as "cryptography." In his or her ability to grasp the relations among apparently quite different domains, the Huehueteco scientist is a cryptographer, that "someone who can at once account for nature and decipher the soul, who can peer into the crannies of matter and read into the folds of the soul." And as an ethicist, the scientist that so peers and reads seeks to produce "creative lines of life" (ibid., 3). It is in this regard that I argue for taking Totonac science seriously as knowledge in the global public interest.

Recent Trauma in Huehuetla

Before proceeding to a set of representative contexts in which presence of mind operates, I need to summarize the recent events that background them. In the process of my research, I was told that in the past (that is, before dependency on the cash economy, before massive natural resource stress, before dependency on rachitic government assistance programs) the present mind was general in Totonac society. Those who speak in this essay (described to me as "our strongest scientists") are among the few Huehuetecos today who can claim presence of mind. A host of ills are attributed to its erosion. The most worrying concern of climatic instability and resource conditions is related to climate change: protracted drought, drying springs, deforestation, and extreme weather events. All of these threaten livelihoods and drive out-migration. In recent years Totonac cultural revitalization initiatives have sought to recover and re-diffuse traditional science precisely to respond to such insecurities. Thus although Huehueteco cosmology does express period-transcendent relationships between natural, human, and divine entities, it is not itself period-transcendent. It is a situated, interested, ethical environmental science, bent to the needs of people negotiating a difficult political, economic, and ecological present (Haraway 1991).

The Huehuetla region has been hard hit by a number of disturbances the past 30 or so years. The most important of these is the demise of coffee developmentalism. Launched by the state in the early 1970s as an agricultural modernization project, Green Revolution coffee monocropping massively transformed Huehuetla's society and environment. Coffee monoculture supplanted subsistence and cash multicropping, intensified the deforestation process already underway in the region, promoted the heavy application of chemical agricultural inputs, and sharply reduced plant diversity. The family-centric coffee production process displaced multifamily cooperative labor systems geared to food production, and full commitment to cash cropping exposed farmers to the vagaries of coffee harvests, markets, and state-sponsored development programs. Also, coffee monocroping has had religious consequences, dislocating local thinking about moral connections among persons, plants, labor, and divinity. There occurred a cosmological and epistemological impoverishment coterminous with reductions in landscapes and production systems. Finally, because it greatly enhanced government intervention in everyday life and in the landscape, the coffee economy consolidated farmer dependency on the state.

The regional coffee economy, however, seems irremediably broken after more than a decade of instability and decline. In 1989, the International Coffee Organization terminated its accords on prices and quotas in the name of market liberation. Prices on the world market abruptly fell more than 30 percent. In that same year, the Salinas de Gortari administration began dissolving the Mexican Coffee Institute, a national assistance program for small coffee producers. That fateful year also saw a killing frost destroy more than half of Huehuetla's coffee fields. The coffee economy never fully recovered from 1989. Most plots in Huehuetla now lie in various states of neglect. The trees are aging and unproductive. Low prices do not cover production costs. Decapitalized farmers are thus slowly but surely abandoning coffee production with no clear alternative in view. The most common resort is to produce more maize in a bid to enhance food security, but maize expands at the expense of tree cover, soil health, and wild plant diversity. In Huehuetla, economic crisis takes its human toll in familiar ways. Communities fray as people turn to cities already crowded with the rural displaced, or more recently to the Mexico–US border. Alcoholism and domestic violence, both serious problems in Huehuetla, increase as the coffee crisis wears on.

Compounding economic trouble with ecological, the incidence of "natural" disaster, both drought and deluge, is on the rise as regional climate becomes more capricious. The Sierra Norte is witnessing a general warming and drying trend which impinges on spring regeneration and dry farming. However, over the past decade, rainfall has become less predictable and more severe in its consequences. Heavier and more protracted showers are falling on an ever-less-stable landscape. One especially destructive storm occurred in 1999. On October 4 and 5 of that year, Tropical Depression 11 passed over the Sierra, dropping torrential rain for sixty consecutive hours. The Huehuetla region was one of the most severely impacted. Entire hillsides were rent in half by the landslides that covered roads and houses with mud and uprooted trees. In the nearby city of Teziutlán, whole neighborhoods were carried by mudslides to canyon floors, killing hundreds. Breaches opened in highway asphalt, bridges washed out, and slides made impassable every major road into Huehuetla for weeks, rendering access to the region extremely difficult. People wandered in the rain, knocking on doors for food or trekking 20 and 30 miles to market towns. Some died of exposure on the way. The state failed miserably to provide effective relief. Local government officials were observed hoarding aid and allocating it to family, friends, and tourists. Some were accused of diverting provisions supplied by the military and/or stealing them from peasant organization warehouses. All political parties used the disaster to swap humanitarian assistance for electoral support.

Indigenous rights mobilization had been on the ascendant in Huehuetla long before October 1999, nourished on the national movement for Indian political autonomy which has become the center of the "ethnic-national question" in Mexico in the wake of the Zapatista uprising in Chiapas (Burguete Cal y Mayor 1999; Díaz-Polanco 1997). Since the mid-1980s, Totonac movements in Huehuetla have provoked an array of reactions by the state that opposed local interests: local government officials have seized communal lands worked by producer organizations heading up rights initiatives; high-profile organization members have suffered scare tactics from stones thrown through windows to death threats; and the Puebla state government has dramatically stepped up state police and military patrols in the municipality.

But after Tropical Depression 11, local organizations shifted focus. Rather than continuing to butt heads with their political adversaries over the distribution of power, they began to explore Totonac self-determination more in terms of the connections among resource bases, community integrity, and God. The storm was regarded as an expression of something deeply wrong in the cosmos. Some imputed the catastrophe to want of faith, with all the social and ecological implications I will spell out later in this chapter. The patron saint, in fact, was said to have abandoned Huehuetla. Thus religious revival, or the reawakening of faith, fell into line with the new conditions and concerns, and keyed the whole disaster response program mounted by organizations.

Community assemblies met to discuss revisiting the Totonac "science of the ancients" to resume bygone communal labor arrangements, envisioning new (more accurately, neotraditional) landscapes. The language they used to describe these new landscapes borrowed from this science of the ancients, the testimony of elders of how it all used to be. And it borrowed, too, from the indigenized Liberation Theology current in the region, from national and international organizations mobilizing for indigenous autonomy, and from globalized sustainability discourses espoused by all parties, from the most radical rights group to the state Secretariat of Rural Development. Families, their food insecurity now thrown into relief, started to doubt the wisdom of exclusive reliance on market crops. Finally, after the appalling behavior of the formal political system, people started to rethink politics altogether, to rethink divisionism within communities, to revitalize community cohesion, and to condemn the national party system that turns even family members against each other. Thus local Totonac organizations began to rethink their priorities; some changed their leadership to better address the issues brought to such a pitch by the catastrophe. This disaster event, then, pushed a substantial readjustment of thought about ethnicity, crops, and territory, the "fit" between culture, landscape, and development autonomy.

Chuchutsipi: Hills, Water and Pueblos

In its present configuration, Huehuetla cosmology reflects a "back to basics" response to weather disturbances like those just summarized. Although this impulse arises out of a specific historical conjuncture, "basics" have to do with a habit of thought that equates to more than any particular politics or response program. Basics require an intelligence, a personal faculty sensitive to the agency of things both human and nonhuman, and to the specific connections among things. The integrity of cosmological components and of the system of their connections depends upon this quality of thought, this cognitive energy.

An analysis of the Totonac concept *chuchutsipi* may help illuminate the character of Totonac connection-making. The word translates as "water-hill" yet means "village." One avenue into the intricacies of *chutchutsipi* as a cosmological pillar is to examine the material and cognitive relations between important springs, which as landscape features are really hill-spring complexes, and communities. Hill/springs are places of fundamental material and religious importance. Hills literally ground Totonac *campesino* farming while springs, as sacred places, are central meeting points between worlds below the ground, water and water divinities, and the people who come into contact with all of these by using the spring. In addition to linking the sacred with agriculture, the hill/spring landscape composite is linked to the idea of the human community. Communities possess certain springs and self-identify thereby, but the connection is richer than that. When I asked people to explain it, replies commonly linked land, water, and community under the rubrics of sacrality and respect, respect for the soil, for life, for God, and among community members.

But respect, one of *chuchutsipi*'s central tenets, is a dual concept. The Totonac idea of respect acknowledges something as valuable and at the same time hazardous. Farmers approach the entire environment in this way in Huehuetla as a degraded environment, a locale vulnerable to climatic vagaries with crisis-prone economies and political instabilities. The non-human is to be cherished, but also recognized as impetuous, ephemeral, sometimes as much violence-prone as nurturing. Uncertainty is built into the cosmology and, in keeping with this duality of respect, hill/springs are fraught with tension, including the communities that use them. Particularly in recent years, community cohesion has fallen victim to conflict between political parties, which divides villages, families and friends.

The Strength of the Hills

Breaking down the hill/spring complex into some of its details will spell out more fully this concept of respect. Hills center a Totonac geography where land, labor, community, and divinity come together. There is little distinction between hills and land in general. Hills, of course, are the material substrate of the chuchutsipi. They shelter the villages and constitute the fields (the forest as well, but deforestation has weakened the forest domain). Hills are the premier environment of Totonac life and labor, primarily agricultural labor, but also plant gathering and woodcutting. There is therefore an essential connection between hills, communities and work. The latter two are in their proper places in the former. At the same time, hills are sites of constant human-divine contact. They are said to form a kind of saints' village supporting human villages, which is to say that people live among the saints. One friend of mine, for instance, says that the hills are houses shared by the Saint Michaels (*sanmigueles*) and whatever other local tutelary powers watch over a community. Hills are therefore the places where many significant elements of the cosmological field, including the weather, come together. Further, by virtue of their mass and altitude, hills are by their nature powerful places, apt sites for communication with the divine, particularly when it comes to requesting protection from climatic vagaries.

The Sensitivities of Water

The nature of water is far more contradictory than that of hills. Springs are perhaps the most "autochthonous" resources insofar as their "owners" (*dueños*, supernatural custodians which every spring has) are almost never Christian figures.[1] But water's

1 There are important exceptions. The original home of patron saint San Salvador is said to be a spring called Kgaxuwachuchut, nearer Huehuetla than Kgoyomachuchut. The saint moved to the latter, reluctantly (always reluctantly do the saints move), after the water at Kgaxuwachuchut had been fouled by careless human use. I spoke with many people about the current "spiritual" identity of springs. Only one said that a saint, Saint Michael, had authority over spring water.

power is a slippery thing, more in the case of springs than other water sources such as rivers and rain. Springs, once the only water sources for subsistence apart from rain catchment and still one of the most important sources, fulfill a human community necessity. Respect for springs, as expressed in quite certain terms within stories of springs told to me by the Totonacs, involves a good measure of anxiety:

> My grandfather cut trees there, near the spring, and one time one of them fell on top of the water. After a while, about three days, he lost his mind. He stripped naked and fled his house. It took them two weeks to find him, just up here on the road. They went to get him with a rope, but by the time they got there, he was fine. And he said, "What happened to me? It's like I was drunk."
>
> He returned to the house and got dressed, then he ate. He ate three platefuls, but the third one he threw against the wall. Then he took off his clothes again and went out. This time they didn't let him get far but tied him with a rope. Then they took him to Ixtepec to a curandero. And the curandero said, "No wonder. The spring is dangerous. You're going to have to take water out of it and bathe him in it, to make amends." But he never was quite the same after that, always a little crazy (Joaquín Esparza Díaz).

And yet the hazards springs present to an absence of mind like that cited in this story have weakened considerably. Many of Huehuetla's springs, in fact, are drying. A Western analyst would probably account for that by noting changes in rainfall patterns and increased runoff due to deforestation and erosion. A Totonac would agree but insist that those issues do not plumb the problem. Ultimately dry-up is a consequence of waning respect for water. People do not care for springs as they once did, nor do they fear them, and the springs themselves have responded to human indifference and carelessness by fading from the cosmology. There is a growing rent in the weave:

> Before, the water was very delicate and nobody messed with it. Where there was a lot of water, you couldn't get near where it was coming out because you would get a fever. Now the people want potable water, but they don't understand that springs are becoming extinct. They don't know how much water there is in the springs, and therefore we're losing the water and we don't see the error. I keep telling people to wash clothes and dishes at the well taps and leave the springs alone, but no one listens. The moment will arrive when there will be no water at all. And now people compete for the water; they want to hoard it. It used to be that you didn't dare dip your foot into a spring, but now it's not like that anymore. The springs are drying up because we have distanced the beings that watch over them. There doesn't exist anymore that supernatural that cares for us. We've lost value and faith. We have to respect each other (Don Joaquín Esparza Díaz).

At least two things stand out in Joaquín's account. First, simply that many springs are disappearing. This is attributable in large degree, again, to changing regional climate patterns. Second, human disrespect for springs in three senses: toward the spring itself, toward the "supernatural"[2] attending it, and toward other community

---

2 Huehuetla Totonacs often refer to non-Christian divinities as "supernaturals" (*sobrenaturales*).

members who rely upon it. All of these infractions have weakened the agency of the water. Weakened water, because it is unable now to chasten the disrespectful, can stem the further deterioration neither of itself nor thought nor the *chuchutsipi*. The presence of mind and the social solidarity supporting the stewardship of resources have frayed.

Conflicts over water, or exercised through water, are frequent in Huehuetla. In particular since party conflicts and ethnic rights movements have polarized pueblos and sometimes families, people sometimes take drastic measures to prejudice someone else's claim to a spring by washing dishes in it, filling it with soap, bathing in it, pouring oil into it, and so on to corrupt the water that will flow to a rival's house. Resources, communities, and the gods suffer for it.

The fate of the springs is deeply troubling for those like Joaquín, "who do not neglect their thoughts" (his words). Absence of mind accounts for the distancing going on between people, water, and the *dueños*. There is always in the presence of mind viewpoint a close connection between the relative power of a substance, the quality of human thought about it, and the divinity responsible for it. If one degenerates, the whole ensemble faces the consequences. Sierra Norte communities have been vulnerable to one or another disorder throughout the decades and centuries, but the sources of disorder have multiplied in recent years. Therefore, presence of mind has never been so important. In its power to order heterogeneous things, places, and times, it bears some resemblance to the "program of topology" of philosopher Michel Serres, one of ANT's precursors and inspirations. For example, about topological space, Serres writes:

> ... the accidents and catastrophes of space. What's closed, what is open, what is a connective path? ...What are the continuous and the discontinuous? What's a threshold, a limit? The elementary program of topology ... a set of connections and junctions to be established—not always already there. The junctions have to be constructed (Serres 1982, 44).

The Huehueteco environment/cosmos is akin to such a space, and presence of mind is like a topological instrument maintaining the junctions and connections (see also Serres 1995). It is important to recall that the latter are not constant throughout history. Today's challenges are different from yesterday's, and yet the "ancient science" that privileges the space-time of the saints and other *dueños* is being reasserted to address present insecurities. Multivectorial in ways inadmissible to Western science, this is still fundamentally a discourse on climate change. It is a way to understand and confront climate change in the face of decreasing rainfall and the drying up of the springs. These changes have fundamental implications in a dry farming economy; they account for increasing environmental (cosmological) insecurity in Huehuetla.

Beyond Huehuetla, the Totonac standpoint should be of interest to scholars attentive to social constructions of climate change and people looking to revitalize and democratize discussions about human responsibility in climate change. The Totonac discourse might be a social construction no less than any other, but it suggests that basic, on-the-ground, personal and interpersonal responsibilities lie at the heart of problems that transcend locality, perhaps to include cosmological problems, problems concerning the universe. Note that Joaquín defines human frailty as heedlessness, bemoans reactivity as opposed to proactivity, and decries absence of mind. The Totonac conception of faith involves mutual respect among people and righteous labor. Faith is virtually synonymous with good thinking, presence of mind itself. We may never witness the welcome of any notion of faith and divinity in hegemonic discourses on global climate change, and I am not suggesting we necessarily should. I do suggest, however, that Totonac understandings of respect, for water, for powers beyond us, for one another, would not be untoward in discussions which rest too heavily on economic priorities set by economic elites, the relative responsibilities of states, and protocols that no one really believes will solve our climate issues. If I know Joaquín, he would be the last to fiddle while the weave of the world becomes ragged. His perspective joins the broader indigenous one referenced by Heather Smith's contribution to this volume, which questions the human-nature dichotomy, the denial of the local, and state-centrism in the conceptual and institutional apparatuses that manage discourses about "global" issues.

Network Theory and Synchronic/Diachronic Cosmology

At this stage of the argument, I want to reintroduce the possibilities opened up by linking Huehueteco presence of mind to network theory. The latter, I believe, helps us understand the social-natural connections sketched so far. It illuminates Huehueteco presence of mind as a form of knowledge and order building, one that contingently weaves an unruly human-nonhuman collective together and binds it toward security. The cosmological field in Huehuetla is clearly a heterogeneous, human-nonhuman collective and thus accords with network theory's bid to demolish sterile oppositions such as subject/object, mind/body, and human/nature. Presence of mind shares something with an approach to knowledge in some of the work of one prominent ANT practitioner, John Law: "… [Knowledge] is the end product of a lot of hard work in which heterogeneous bits and pieces ... that would like to make off on their own are juxtaposed into a patterned network which overcomes their resistance" (Law 2002, 2). Here Law agrees with Latour when the latter argues that the mind does not create the world as it pleases but "is seized, modified, altered, possessed by non-humans who, in their turn…alter their trajectories, destinies, histories" (Latour 1999, 282). Law's and Latour's use of the passive voice ("are juxtaposed" and "seized") is telling, because it indicates a reluctance to entertain human intentionality in piecing together collectives.[3] This is, of course, faithful to the equality between the human and nonhuman in networks that ANT insists upon. Hetherington and Law, for instance, define ANT as a "relational approach that does not recognize the distinction between humans and nonhumans ..." (2000, 127).

But just what is it that juxtaposes heterogeneous bits and pieces? In its turn, how does thought engage, not to say order, the non-human and human alike? How does

3 See Arce and Long 2000, Golinski 1998, and Lynch 1996 for critiques of ANT along similar lines.

it bring and hold together a complex construction like *chuchutsipi*, or "climate," in any discursive tradition (Demeritt 2001)? "Topology," which bridges distances by bringing heterogeneous elements together, is firmly positioned in the vocabulary of ANT (see Mol and Law 1994); however, ANT cannot do justice to the topological mind I find in Huehueteco cryptographers. In a world more and more prone to decay (a kind of post-coffee entropy), presence of mind works negentropically, drawing essential material and moral connections among landscapes, people and gods. It tries to sustain all of these in viable relationships. This is an effect of drawing the cosmology together in the first place, which is what presence of mind as a knowledge does, never, however, once and for all. It is always a working knowledge both in the sense of knowledge-in-labor and knowledge-never-complete.

"Timeless" relations do subsist among soil, labor, plants, and saints; the call for a return to tradition would be meaningless otherwise. Yet the cosmology is far from static, responding today to a bundle of erosive processes: resource bases, economies, and social solidarities all have weakened; regional drying and more frequent flooding menace the Sierra.[4] Presence of mind is thus fully historical. It ranges social, ecological and celestial fields, mixing spaces and times, trying to piece together orders from conjunctures of disorder. The mind in Huehuetla cosmology is indeed worked on by a host of nonhumans; this is a mind-in-a-body impinged on by the world. Saints and water and trees and communities are entities with "lives of their own", trajectories that resist human control; they sometimes flout the ensemble. The work of presence of mind and respect as a concept set and a practice is to overcome these kinds of resistances in the interests of security. The thought that manages respect for the world aims at what ecologists call "system resilience" (Adger 2000; Birkes and Folke 1998; Holling and Sanderson 1996) against dismayingly normal hazards. It is in this regard intentional.

And thus it is *political*. One should never forget that the pursuit of presence of mind today is occurring in the context of Totonac mobilization for self-determination. Strictly political questions aside, climate change is one of the greatest threats to a Totonac-determined future. The cosmology defines a moral geography, a human-nonhuman sociality, an environmental ethic in the most comprehensive terms. Presence of mind constructs a complex heterogeneous ensemble like *chuchutsipi* and makes that ensemble available to a political thought and practice in opposition to current political economy and (even if people like Joaquin are not aware of it) to the globalism of dominant discourses on climate change and what we should do about it. This is, after all, about power, more specifically about what Isabelle Stengers calls "cosmopolitics." Latour paraphrases Stengers this way: "Yes, we live in a hybrid world made up at once of gods, people, stars, electrons, nuclear plants, and markets,

4 See Bunker 1985 for a thermodynamic approach to development history in Amazonia. While coffee developmentalism in Huehuetla is dissimilar in many ways, the processes of environmental decay, from low to high entropy, it set in motion may be compared to the extractive economies Bunker analyzes (see chapter 3).

and it is our duty to turn it into either an 'unruly shambles' or an 'ordered whole' ..." (1999, 16).

Presence of mind reflects such a duty. ANT's reluctance to privilege the human notwithstanding, Totonac cosmology demonstrates that human thought animates relations and dynamics among the elements of an ensemble, and for certain purposes. Huehueteco science brings a cosmology to life for cosmopolitical reasons. There is something in it, namely its ability to grant nonhuman entities subjectivity (not mere agency, there is a difference) and its bid to keep a capricious cosmos in good working order that exceeds ANT's vision of topology.

Plants, Thought, Faith and Work

To firm up this argument, I wish now to flesh out the implications of climate change for Huehueteco agriculture, the primary site of righteous labor. While this theme may provide little direct commentary on climate for non-Totonacs, Huehuetecos are well aware of the relations among deforestation, disruptions in ecological relationships between trees and other plants, changes in air and soil temperature, and plant diversity reduction. Ultimately, climate is the issue, but it is crucial to bear in mind that the disappearance of plants amounts to the disappearance of subjects in the human-nonhuman collective. According to the doctrines of presence of mind, the righteous individual working with plants seeks to attune his or her thought with the thought of the plant, which is a voice of God. The implications are ominous; agricultural reductionism means silences in the dialogue among actors in the cosmology, further gaping holes in the weave of the world.

Food for Thought/Faith and Thought/Faith for Food

Huehuetecos recognize the necessity, the goodness and the wisdom of food production, appreciate the connection between growing food and maintaining communities in fellowship with the land. The key to maintaining a vital relationship between agriculture and thought is food crop diversity. The variety of food plants (both crops and plants gathered wild) manifests in almost Thomist fashion the manifoldness of God's presence in the world. Looking upon, dwelling upon, working with and consuming this multiplicity enhances the connection between mind, body, the land and God. I was told on many occasions that patron saints "order" people to produce a variety of food crops. But as subsistence systems have forfeited ground to an agricultural consciousness geared almost exclusively to coffee and corn, as deforestation alters cloud forest ecosystems, and heavy rains wash away more and more topsoil, plant diversity has fallen out of mind. Even maize crops are chronically vulnerable to drought, deluge and the burning winds that visit Huehuetla in April and May.

In Totonac thought, these conditions are inseparable both from properties and acts of God and the quality of relations between people within families, within pueblos, and among Totonacs. Food security through subsistence farming requires humility, goodwill, hard work, social cooperation and concord, and steady propitiation.[5] These are the virtues and the practices without which harvests fall to catastrophes. Labor participates here mainly through subsistence agricultural work, especially if that work is communal. Cooperative labor among friends and kin during maize sowing is all that remains today of an ample labor reciprocity regime once involving vegetable planting, weeding and fertilizing, harvesting and house building. *La mano vuelta*, as the institution is called in Spanish, combines practical concerns,[6] social cohesion, spiritual satisfaction and sacred imperatives. It brings men and women together to work even private land in common. *Mano vuelta* does involve a gendered division of labor: men sow while women prepare food. But here food preparation is ritualized as the essential complement to the food production of sowing, and *mano vuelta* is one of the few social contexts that explicitly acknowledge women's labor as equal in value to men's.

One day I asked Joaquín how the switch in the early 1970s to coffee monocropping changed people's thinking about God. He replied, "People stopped thinking about the plants. They stopped thinking about God. Why do you need to think about God when you have money?" Joaquín consistently stresses the need to restore faith, which is another way of terming presence of mind. He says, specifically, "We need to give faith to our harvests." By saying *giving* faith to our harvests, not having it *in* our harvests, Joaquin speaks as if faith produced by presence of mind were the life force necessary to animate the plant. He finds the present age wanting in comparison with a past one in which human–plant relations enjoyed a higher degree of immediacy and intimacy. There was less need to give faith then; faith was immanent in the plant–person–God nexus.

As plants have fallen to sparer and sparer agricultural schemes and climate change makes their return all the more unlikely, likewise presence of mind has diminished. When one ceases thinking about the plants, or, which amounts to the same thing, thinking about God through work with plants, one loses the faith linking humans, plants, labor and the divine. The loss of thought on the God/plant (the loss of thought in general considering this Totonac model of thought) has deep spiritual and moral, as well as practical consequences. Totonac scientists characterize the present era as one of faded capacities, specifically of faithless work and feeble thought. Human potentials are dwindling alongside the capacities of the land and the plants.

5 I should note that "subsistence farming" and "food security" do not preclude the buying and selling of food crops. No one romanticizes the self-sufficient indigenous peasant growing food in a world innocent of money, and farmers have always sold their surpluses in Huehuetla's markets and bought food from other farmers. But a cash economy centered on local food production is conceived quite differently from one dependent on cash monocropping.

6 Ten men plant a corn field a great deal faster than one, and given the frequency of inclement weather in Huehuetla, it can be crucial to plant in one or two days.

Climate Change, Totonac Scientists, and Us

I have argued that regional climatic drying, although not the only climatic issue Huehuetecos face, is the gravest ecological threat to Totonacs as they struggle to rebuild resource bases, economies and communities for a self-determined future. Their struggles with climatic issues bear, therefore, on struggles for cultural, political, and economic autonomy. These are issues with a reach far beyond the Sierra Norte. Indigeneity in identity politics and debates in international law has been on the rise since the 1970s. I have conceded that Totonac conceptions of climate may be as constructed as any other, but this is merely to recall that there is a politics to social constructions which to this point, in neither scholarship nor policy discussions, has been sufficiently democratic. There has indeed been a clamor to get "the indigenous voice" into climate change debates, but that clamor, I would argue, has been largely ignored in the "high level" talks about confronting the perils of climate change, just as it has been marginalized from "the global" in Western (or Northern)-dominated visions of the nature of these problems. In other words, the "green governmentality" of global environmental management (Bäckstrand and Lövbrand this volume) provides little but lip service to indigenous approaches to problem-solving. The multicultural character of our time requires that we expand the space of intersubjectivity in the dialogue.

Specifically, unlike the approach to climate change that defines the problem as global, addresses it in terms of emissions, targets, timetables, and states, and which distances *concrete* human responsibility from nature (see again Heather Smith this volume), Huehueteco cryptographers view their climate problems as eminently local and social, not managerial. As post-NAFTA Mexicans, Totonacs are well aware that they stand in a position of weakness vis-à-vis the much more resource-intensive United States. As Indian coffee farmers, they know they have been swept perforce into Mexican modernization schemes. They know that their landscapes and their weather register the consequences of priorities not always their own. They know that their poverty takes less out of the planet's life support systems than Western wealth. Why, then, do they look first to themselves in identifying problems like climate change in the Sierra Norte and prescribing remedies? I would suggest that the "parochialism" of the Totonac purview grounds Totonac praxis in such a way that local people may draw upon local human capacities to deal with local issues. Their approach arises from a sense not of powerlessness but of the solidity of situated knowledge (Haraway 1991). Although cosmopolitics is part and parcel of global discussions about indigenous rights and environmentalism, actual problem-solving is most effectively pursued where, in this case, the traditional environmental knowledge that Heather Smith mentions (this volume) can address the complexities of environmental issues in place. In other words actions should be pursued where local scientists have a grasp of the physical and social terrain, and can look into the creases of matter and the folds of the soul to discern ways in which lived problems are related.

How might we look to people like Joaquín to rethink our own responsibilities to the human community, however we make that abstraction take on flesh in our own circumstances? How do we clarify our responsibilities to climate itself? I have suggested that one current of intellectual resources already available in the North, namely Science and Technology Studies, network theory, and particularly ANT, may help us grasp both Totonac science and some gaps in our own. ANT, like Huehueteco cosmology, plays down the human-nature dualism and grants that orders are built by the agency of nonhumans as well as humans. I have also argued, however, that ANT, in its reluctance to grant any degree of human exceptionalism, fails to accommodate my understanding of Totonac presence of mind, a kind of cosmological pivot that links the constructive faculties of water, soil, plants, people and saints. This is what distinguishes Native American theory and methodology from even the most radical Northern framework. While I would not insist that we all must adopt the spiritual-ecological worldview that marks Totonac cosmology, we should realize that climate amounts to more than statistics, carbon sinks, pollution credits, and emissions quotas. Joaquín knows that climate issues are social issues, embedded in a field of social relations that involves human-to-human interactions yet exceeds them. There is more to climate change than concerns about human welfare, and I am not at all certain that every consideration regarded as "spiritual" should be barred from serious discussions about climate change. Further, Huehueteco measures to restore the world (calibrated to the local, present-to-the-senses, vividly intuited weave of the world) may be a refreshing, not to say necessary, alternative to nation-states as greenhouse gas emitters, to governments impotently confronting the big polluters, and to sacrifice after sacrifice on the altar of economic growth.

Totonac "righteous" labor, essential economic activity is a kinetic expression of the present mind that effects constructive material relationships among cosmological domains. The economic choices one makes are always already ecological and therefore bear on climate, among other things. Specifically, Totonacs, as did Marx, recognize that labor drives the metabolism of nature and society; labor lies at the heart of what it means to be human. Labor, apart from being a crucial context of human relationships, is one way in which we extend the body into the matrix of nature (White 1995). Labor is a fundamental site of interaction between human and nonhuman. The Totonac vision of labor has the virtue of linking labor to mindfulness. We should think through our labor, to gloss one important Totonac precept.

But I would like to conclude with a last note on the politics of science, that is, on cosmopolitics. In an essay on enriching analysis of scientific practice, Andrew Pickering urges us to "think of conceptual structures as ... located in fields of agency, and of the transformation and extension of such structures as emerging in dialectics of resistance and accommodation within those fields ..." (1995, 418). This chapter has attempted to follow Pickering's suggestion by attending to the historicity and the politics of concepts, and by delineating the diverse field of environmental agency in Huehuetla. Latour argues that "It is only through an extraordinary shrinking of politics that it has been restricted to the values, interests, opinions, and social forces of isolated, naked humans" (1999, 290).

Dominant northern discourses on climate change regrettably reflect this shrinking that Totonac discourses do not. And cosmological though they be, they are as politically motivated as any, perhaps more than most given the political, economic and ecological straits in which Totonacs find themselves today. These are immediate life and death issues for them, not theoretical discourses. Far from compromising Totonac climate change discourse (or any other) by pointing out its interested constructedness, not to mention its exoticism from a Western scientific standpoint, we need to honor the heteroglossic field (Bakhtin 1981) of scientific definitions of the world by inviting people like Huehueteco scientists to the debate table. If the Kyoto Protocol is any indication, a Northern-dominated, state-centric strategy will not turn the tide of climate change.

The Huehuetecos I have represented here are not crotchety old men, dinosaurs in a Third World backwater whom "science" has outpaced. Joaquín, for instance, is in his forties. Nor are they noble savages (Joaquín would in particular be much amused at that idea) but quite practical men. Young and old, all are "traditional" and innovative simultaneously, free thinkers about thought, and able to process lucidly the consequences of history and the prospects for the future. They are deeply concerned about thought loss, or the fragmentation of thought, as Huehuetecos have converted their lands to coffee monocultures, become virtual wards of the state, and collapsed into blind feuds between political parties.

These men are frustrated about the fate of knowledge and thought in their community. Social and environmental changes have buried the right patterns of thinking and understanding that "everyone knows" or should know. And thought (the force of right knowledge) has been fragmented into solipsistic consciousnesses, become rootless or debased by the atomization of Huehuetla society. This has proceeded in tandem with, and has contributed to, Huehuetla's deepening insecurity: the deterioration of its environment, the worsening economic poverty of its people, the hostilities that prejudice communities, and the increasing incidence of natural disaster. These are all consequences of the absent mind, but far from throwing up their hands or leaving it to governments to confront their problems, Huehueteco scientists are trying to do something about it. They want to see things repaired, rewoven, and that will require present minds and a cosmopolitics equal to the stresses of the here and now. This is a question of power in several senses (see Wolf 1990).

Huehuetla cosmology exceeds the sum of its parts in that it expresses not simply how things are in the universe, but how they should or must be. Cosmology is not just about order, it is about *ordering*. This ordering intends security, not just for humans but for all subjects in the collective. Environmental security, livelihood security, social sustainability, and political sustainability, all of these fall under the rubric "security" as presence of mind pursues it. This is a discourse on climate that others might consider more seriously. Building on what has been raised in previous chapters by considering the disconnect now present in climate change discourses between the local and the global, a serious consideration of alternative perspectives such as Huehuetla cosmology is in order.

References

Adger, W.N. (2000), Social and ecological resilience: are they related? *Progress in Human Geography*, 24, pp. 347–364.

Allen, C.J. (1988), *The Hold Life Has*. Washington: Smithsonian Institution Press.

Arce, A. and Long, N. (2000), "Reconfiguring modernity and development from an anthropological perspective" in A. Arce and N. Long, eds. *Anthropology, Development, and Modernities*. London and New York: Routledge, pp. 1–31.

Bakhtin, M. (1981), *The Dialogic Imagination*. Austin: University of Texas Press.

Birkes, F. and Folke, C., eds. (1998), *Linking Social and Ecological Systems*. Cambridge: Cambridge University Press.

Bunker, S.G. (1985), *Underdeveloping the Amazon*. Urbana and Chicago: University of Illinois Press.

Burguete Cal y Mayor, A., ed. (1999), *Experiencias de autonomía indígena*. Copenhagen: IWGIA Documento 28.

Deleuze, G. (1993), *The Fold*. Minneapolis: University of Minnesota Press.

Deleuze, G. and Guattari, F. (1987), *A Thousand Plateaus*. Minneapolis: University of Minnesota Press.

Demeritt, D. (2001), The Construction of Global Warming and the Politics of Science. *Annals of the Association of American Geographers*, 91, pp. 307–338.

Descola, P. (1996), *In the Society of Nature: A Native Ecology of Amazonia*. Cambridge: Cambridge University Press.

Diaz-Polanco, H. (1997), *La Rebelion Zapatista y la Autonomia*. Mexico: Siglo XXI.

Golinski, J. (1998), *Making Natural Knowledge: Constructivism and the History of Science*. Cambridge: Cambridge University Press.

Haraway, D. (1991), *Simians, Cyborgs, and Women*. New York and London: Routledge.

Hetherington, K. and Law, J. (2000), After networks. *Environment and Planning D: Society and Space*, 18, pp. 127–132.

Holling, C.S. and Sanderson, S. (1996), "Dynamics of (Dis)harmony in Ecological and Social Systems" in S. Hanna, C. Folke and K. Goran-Maler, eds. *Rights to Nature*. Washington, D.C.: Island Press, pp. 57–86.

Kuletz, V. (1998), *The Tainted Desert: Environmental and Social Ruin in the American West*. London and New York: Routledge.

Latour, B. (2005), *Reassembling the Social: An Introduction of Actor-Network-Theory*. New York: Oxford University Press.

________. (1999), *Pandora's Hope*. Cambridge: Harvard University Press.

________.. (1996), *Aramis or the Love of Technology*. Cambridge, MA: Harvard University Press.

________. (1993), *We Have Never Been Modern*. Cambridge, MA: Harvard University Press.

Law, J. (2002; first published 1992), *Notes on the Theory of the Actor Network: Ordering, Strategy and Heterogeneity*. < http://www.lancs.ac.uk/fss/sociology/papers/law-notes-on-ant.pdf>.

Law, J. and Hassard, J., eds. (1999), *Actor Network Theory and after*. Oxford: Blackwell Publishers/The Sociological Review.

Lee, N. and Brown, S. (1994), Otherness and the Actor Network. *American Behavioral Scientist*, 37, pp. 772–790.

Lynch, M. (1996), DeKanting agency: comments on Bruno Latour's "On Interobjectivity". *Mind, Culture, and Activity*, 3, pp. 246–51.

Mol, A. and Law, J. (1994), Regions, Networks and Fluids: Anaemia and Social Topology. *Social Studies of Science*, 24, pp. 641–671.

Murdoch, J. (2001), Ecologising Sociology: Actor-Network Theory, Co-construction and the Problem of Human Exemptionalism. *Sociology*, 35, pp. 111–133.

Pickering, A. (1995), Concepts and the Mangle of Practice: Constructing Quaternions. *South Atlantic Quarterly*, 94, pp. 417–465.

Serres, M. (1982), *Hermes*. Baltimore: The Johns Hopkins University Press.

________. (1991), *Rome: The Book of Foundations*. Stanford: Stanford University Press.

Serres, M. with Latour, B. (1995), *Conversations on Science, Culture and Time*. Ann Arbor: University of Michigan Press.

Whatmore, S. (2001), *Hybrid Geographies: Natures Cultures Spaces*. London: Sage.

White, R. (1995), "Are You an Environmentalist or Do You Work for a Living?": Work and Nature in W. Cronon, ed. *Uncommon Ground*. New York: Norton, pp. 171–185.

Wolf, E.R. (1990), Distinguished Lecture: Facing Power – Old Insights, New Questions. *American Anthropologist*, 92, pp. 586–596.

Chapter 11

Conclusion: The Constructions of Climate Change

Loren R. Cass and Mary E. Pettenger

A myth is a story that has validity based on its social acceptance and worldview, but also has questionable authenticity as its legitimacy is often unverifiable. A fact is something that exists as an objective reality, as something undeniable (at least until proven otherwise). Somewhere between the two lie climate change and its social interpretations. One example illustrates the contested understandings of climate change and sets the stage for important themes that resonate in this book. Articles materialize almost daily that document or dispute the existence (facts) of climate change. A clear example is the debate linking hurricanes/typhoons and global warming, especially in light of the 2005 Atlantic hurricane and 2006 Pacific typhoon seasons. As an August 2006 article discusses, academics have "published a flurry of papers either supporting or debunking the idea that warmer temperatures linked to human activity are fueling more intense storms." Importantly when one thinks of myths versus facts, the article notes that "Both sides are using identical data but coming up with conflicting conclusions" (Eilperin 2006). What social reality is being constructed when facts are interpreted into myths potentially by "both sides"? How is knowledge forming or being controlled by power? This becomes more salient when policy makers are left to decide which report to believe, and the public is left with competing scientific interpretations, uncertainty, and half-truths. What does it mean to say that climate change is occurring? What voices are privileged in the discussion, and what voices are marginalized? How does policy relate to underlying discursive and normative debates? These are some of the questions addressed in the chapters, and we return to these questions in the conclusion.

Categorizing Climate Change

In our own process of social construction, we are ordering our interpretations of this book by creating categories (much like Paterson and Stripple discuss in the introduction to Chapter 6). As such, we have constructed several perspectives, or rather drawn out what we deem to be important for the reader. The primary goal of this book is to begin the process of understanding climate change politics through the lens of constructivism. In the introduction we noted that powerful actors, norms and discourses have profoundly shaped the process of climate change politics. We turn now to identifying these actors and their roles, and the processes in which dominant norms and discourses, as well as new voices, emerge. We assume, as constructivism does, that change is possible (and potentially inevitable), and that these alternative voices may become more prominent, or perhaps absorbed into the prevailing norms, discourses and policies in the future.

We have divided the following section into two parts. Through our own processes of exploring the social construction of climate change, we have discovered and synthesized two main themes with several sub-themes emerging from the compilation of the chapters. The first theme relates to the processes of change in the social construction of climate change. The chapters, in their emphasis on norms and discourses, highlight the disputed nature of the origins and maintenance of the dominant norms and discourses embedded in climate change policy, as well as delineate the alternative norms and discourses that have been marginalized. One important question to be explored is how dominant norms and discourses gain prominence. Who or what enables this process? In addition, how do marginalized norms and discourses challenge the dominant forces, and perhaps gain saliency?

A second set of themes relates to several sources of change presented in the chapters, including (1) nature/human interaction, (2) construction of knowledge through science, (3) the social role of humans, and (4) participation as a vector of change. We now turn to further exploration of these points and the substance of the chapters.

Processes of Change—Discourses and Norms

The divide in the constructivist approaches represented in this volume reflects the diversity within constructivism in general. By consciously focusing on discourse and norm-based approaches, the chapters highlight distinct and important elements of the social construction of climate change. Three points of emphasis related to the processes of normative and discursive change emerge from the chapters and suggest material for continued discussions between the various constructivist approaches. It is not the objective here to try to bridge the constructivist approaches; rather, we raise these points to contribute to the ongoing discussion within constructivism. We concur in the recent call by Klotz and Lynch for "constructivists of all persuasions to recognize our shared aim to produce scholarship that carefully uses and evaluates evidence. If we lose sight of our commonalities, we miss an opportunity to learn from each other about how to understand *and* explain international relations" (2006, 362).

The first point that flows from the chapters is that the dominant discourses surrounding the politics of climate change play a critical role in privileging particular actors, problem definitions, and solutions in the policy process. While such a conclusion appears obvious and unremarkable, much of the scholarship on climate change does not take seriously the ways in which underlying discourses and norms dictate the framing of the problem and its potential solutions. Only in highlighting these discourses and norms is it possible to understand the construction of climate change as a problem and explain or interpret the evolution of the political response to it. For example, Paterson and Stripple discuss one of the most basic discourses of international relations (territoriality and sovereignty) and demonstrate how "territorialization" creates the fundamental framework for the construction of climate change and its associated discursive/normative disputes. As Paterson and Stripple note, related to the issue of politics of scale, "Climate change policy in this context is fundamentally constructed through the twin lenses of national security and national economic strategy in a globalizing economy. In both, the state is fundamentally the 'master discourse,' which serves to legitimate other discourses" (p. 154).

The discourses of territoriality and sovereignty and the associated framing of climate change as a global problem requiring a response at the level of states immediately privileges certain discourses and potential solutions while marginalizing alternatives. Heather Smith notes that "The focus on 'the state' limits our field of vision" (p. 199). Heather Smith and William Smith both demonstrate how the very definition of climate change as global and state-centered marginalizes indigenous peoples and the alternative discourses that they present. As Heather Smith notes, "In the case of climate change we find that it is the voices of the elites, states, and sometimes regions that are typically represented in multilateral international negotiations, while indigenous peoples and their counterhegemonic discourses are masked and marginalized" (p. 200). Lahsen and William Smith also emphasize that some leaders of developing states are distrustful of developed states' conceptualization of the problem and potential responses, viewing discussions of "global" solutions actually as "developed state" solutions and, inevitably, neocolonialism. Consequently, existing climate policies flow out of the dominant discourses and the associated construction of the problem. The obvious conclusion is that a dramatic change in the course of climate policy requires a shift in the underlying discourses. As Paterson and Stripple note, "The question is thus less 'how can we get states beyond their parochial concerns?' and more 'what precise form of discursive construction of territoriality and its alternatives permit and constrain the development of climate politics?" (p. 164).

While territoriality and sovereignty provide the central discourse for climate policy, Bäckstrand and Lövbrand identify the more specific discourses that characterize climate change. They argue that three alternative discourses, "Green Governmentality," "Ecological Modernization" and "Civic Environmentalism," currently structure environmental politics generally and climate policy in particular. The first two discourses are consistent with the larger discourse of territoriality, while the final discourse challenges it. The fact that Civic Environmentalism is not fully consistent with the underlying discourse of territoriality explains in part why it has made the least progress in becoming established among the dominant discourses. It also suggests that there are substantial obstacles for Civic Environmentalism to achieve political salience. However, Pettenger notes that the Civic Environmentalism discourse has achieved greater relevance in the Netherlands, which may provide an opportunity to explore the forces producing discursive shifts within a state over time. Implicit in Fogel's discussion of American climate policy is a similar recognition of discursive change over time. The chapters emphasize the importance of identifying the dominant discourses and the interactions among them to understand why certain actors, problem definitions, and solutions are privileged in the policy process.

The second point of emphasis that emerges from the chapters is that there are significant connections between norms and discourses that have not been sufficiently examined in the constructivist empirical literature. There is a divide between positivist and post-positivist approaches to constructivism that is not easily breached (Checkel 2006, Dunn 2006). However, the chapters suggest that each approach provides potentially complementary insights into the social construction of climate change. International norms provide concrete expression to the underlying discourses. For example, norms contain the specific "oughts" and "ought-nots" that often flow logically from the underlying discourses. It is thus necessary to study the evolution of norms within their discursive contexts. For example, the competing Green Governmentality and Ecological Modernization discourses are at the heart of the conflicts between economic efficiency and national responsibility norms that Pettenger, Fogel, Hattori, and Cass address in their chapters, linking the legitimacy of the competing norms with their discursive justifications. In this linked process, the normative debates are closely linked to domestic policy struggles and are more closely aligned with concrete policy decisions than the discursive struggles. However, the normative debates occur within the broader discursive framing of the problem, and the potential policy solutions flow out of the underlying discourses, the associated norms, and ultimately determine the policy processes.

Fogel in particular emphasizes the connections among discourses, norms, and policy. She notes that in the American context international norms required a reframing of the American discourses on climate change to permit the norms to be viewed as legitimate domestically. She suggests that though American national policy has not yet changed, the domestic discourses surrounding climate change appear to be shifting in ways that may lead to the increased salience of international norms and ultimately to reforms in American policy that will reflect the changes in the dominant discourses and associated norms. Pettenger also emphasizes the importance of the relationships between discourses and norms in the evolution of the equity norm and the reform civic environmentalism discourse in climate policy in the Netherlands. In sum, there would appear to be some useful links across the discourse and norms literatures that could be mutually beneficial in providing greater depth to understanding how discourses and norms change, and concretely how those changes are reflected in the dominant norms and ultimately the policies pursued.

The third point of emphasis that emerges from the chapters in this volume is that agency poses a challenge for both the norms and discourse approaches. The agent-structure debate is a major point of contention among constructivist scholars. It is analytically easier to describe the underlying discourses and normative debates than it is to understand how the dominant discourse has emerged or to explain why one norm achieves political salience and a competing norm does not. Heather Smith identifies some of these critical questions in her chapter, "from a critical perspective we are encouraged to ask for whom? For whom is the discourse? Who has constructed it and what interests does it serve to protect and promote? Who is excluded?" (pp. 199) These are the essential questions if we are to begin to address the questions of change in norms and discourses over time. If we want to understand the interplay of power and knowledge, we need to know who has power and how they use it, and who is shaping and mobilizing knowledge.

The strength of the discourse chapters in this volume is the articulation of the competition between dominant and alternative discourses that shape the construction of climate change as a political problem. However, the descriptions cannot explain why these particular discourses have emerged and why one discourse is more influential than alternative discourses (nor do some who adopt the discourse perspective ask such questions). For example, Bäckstrand and Lövbrand assert, "... through the adoption of the Kyoto Protocol the radical equity agenda in the civic environmentalism discourse has been gradually marginalized in favor of liberal narratives of ecological modernization" (p. 136). They do begin to point to some dominant agents in their discussions of "discourse coalitions," "knowledge brokers," the US, the European Union and non-governmental organizations (NGOs). Heather Smith and William Smith also speak of the marginalization of indigenous peoples, and Lahsen notes the exclusion of developing country scientists. Who are the essential actors who have marginalized the alternative narratives, and how have they accomplished this outcome? What determines their relative influence over the dominant norms/discourses? What changes would need to occur for an alternative narrative to achieve greater influence? These are some of the essential questions that remain to be addressed to achieve a deeper understanding of the role of agency in the evolution of dominant discourses.

The norms chapters focus more explicitly on agency and power relationships, but the chapters also struggle to explain why one norm emerges as more influential than another. Pettenger, Hattori, Fogel, and Cass focus on agency and the role of power and influence in the development of climate policy in the Netherlands, Japan, the United States, Germany, and the United Kingdom. Each chapter addresses domestic normative debates that are tied to underlying discursive foundations, which legitimate the competing normative claims. For example, Fogel identifies the critical actors within the American context that have shaped the US response to emergent international norms. She also illustrates how actors draw upon the underlying discourses to legitimate their policy preferences. The norm-centered chapters also highlight the role of power and knowledge in framing climate change for domestic political debate. Scientists played critical roles in framing climate change through the introduction and interpretation of scientific evidence. Norm entrepreneurs sought to frame climate change to reflect their beliefs and policy prescriptions. Political leaders actively sought to shape the international negotiations to meet domestic political objectives.

These chapters raise important questions about the role of agency in shaping discursive and normative debates. What is less clear in these chapters is why the public and policy makers accept some norms and reject others. Material interests play a role for some actors, the perception of legitimacy and fairness is clearly significant, and the congruence between international norms and preexisting domestic political norms and discourses is also relevant. However, there are many questions related to agency and domestic norm salience that deserve further attention in future research. These points of emphasis focus on the process of norm/discourse change. We now shift to identifying areas where change may emerge.

Sources of Change

The focal point of this book is to ascertain the underlying processes of social construction of climate change. This issue is germane as well to the field of International Relations as we are often challenged to explain change, for example, note the extensive debate surrounding the end of the Cold War. This book asks its own questions related to change, such as why did the US withdraw from the Kyoto Protocol? How has ecological modernization risen to prominence? What we see from the previous section, and the chapters themselves, is that change was undoubtedly occurring exogenously to climate change agents. Physical realities were fluctuating (hotter weather, more destructive storms, etc.) while agents in their social contexts struggled to interpret, categorize and respond to these changes. The emergence of discourses and norms further defined this external context, formed through social interactions between those touched by climate change in all its myriad material realities and social interpretations. In this section, we suggest several pathways of change that may shift the social construction of climate change and future climate change policy.

Nature/Human Interaction The first opportunity for change that we found in perusing the chapters is the interaction between nature and humans. Climate in this sense is considered part of nature. Thus, how nature is perceived is integral to understanding how climate change policy is constructed. Nature for some is an external force to be manipulated and controlled (as seen in Pettenger's example of the Dutch holding back the sea and gaining territory). As such climate change becomes an issue of modernism, requiring human ingenuity, science and technology to control (present in Bäckstrand and Lövbrand's overview of the green governmentality discourse, and the Japanese adoption of energy efficiency by Hattori). For others, nature is a product to be consumed (ecological modernization in Bäckstrand and Lövbrand, the US economic position in Cass, and economic growth in Japan as presented by Hattori) or is seen as a tool for domination (as seen in Paterson and Stripple's discussion of sinks creating northern neo-colonialism in the south, and the ravages of globalization on local scales presented by William Smith). As these chapters convey, climate change may never be adequately addressed as long as humans refuse to radically change their socially constructed understandings and continue to use economics and power to dominate nature.

There are gray spaces in this critique however. Both the Dutch and Japanese cases illustrate self-identities that view nature from a stewardship role, as something to be protected (yet still dominated by humans). This norm is also present in Fogel's reflections on the US, and Cass's discussion of German and UK policy. In addition, several chapters illuminate alternative understandings of nature that may bring forth new insights to address climate change. For example, Heather Smith and William Smith present indigenous voices that view humans as intrinsically part of nature, rather than as controllers of nature. From this perspective, nature takes on its own inherent value and agency, sans human consumption. Nature, as William Smith notes, is both nurturing and punishing. It provides for humans and produces our external/material realities, including sea level rise, drought and violent storms.

Additionally, Heather Smith and William Smith mention the role of the sacred, of nature as a spiritual force that guides and constrains human actions. While neither suggest that we should adopt a faith-based approach to climate change, the notion of re-conceptualizing nature is relevant to understanding the social construction of climate change. If we perceive climate, as part of nature, as both agent and structure, we might form new understandings of how to respect and co-exist with this force. Such thoughts are not completely alien from the discourses and normative debates in climate change as illustrated by Fogel. She outlines the rise of religious organizations in the US promoting progressive climate policies based on conceptions of faith and nature, for example, the "What would Jesus drive?" campaign.

Thus, nature is portrayed in the book on one hand as a static force to be manipulated and consumed. Conversely we could also view nature as its own agent and structure, with its own essence, while at the same time appearing as a structure, external to humans, the ultimate other shaping human life.[1] From this perspective, perhaps climate change policy might emerge that completely halts the emissions of greenhouse gases for the sake of nature, rather than for human survival or economic benefits. In sum, the struggle to define the relationship between humans and nature continues, perhaps leading to radical new conceptions or to the maintenance of the status quo with the potential for nature to emerge as executioner.

Knowledge Creation and Science The second opportunity for change is in the role of science in constructing climate change knowledge. As discussed previously and throughout the book, what is sought is irrefutable evidence that climate change exists. Yet, as seen in the article discussed at the beginning of this chapter, science relies on objective facts which must be interpreted within political contexts where objectivity is often negated. Several chapters explicitly address this issue. Bäckstrand and Lövbrand detail the principles of the green governmentality discourse based upon the construction of knowledge by science (albeit Northern generated science). In this discourse science is a tool that views nature as a large machine whose sole

1 One wonders then in this context if we should embrace, rather than reject Schlesinger's (2005) stinging commentary on climate change as a new theology. Perhaps if we view nature as both a sustainer and a destroyer ("the force of nature"), we may see nature as having divine properties and as such we could perceive its role as judge and warden of humans for destroying nature.

purpose is to support human life. Science becomes the means to confront and abate climate change for human survival.

For others, as Lahsen discusses in Chapter 7, science has become politics by another means (this perspective is mirrored as well in Paterson and Stripple's discussion of sinks). Science is "situated knowledge and a potential vector for hegemonic power" (Lahsen, p. 186) As she notes, future efforts that attempt to overcome the reluctance of decision makers to make costly changes to address climate change, based on the formation of knowledge grounded in scientific facts, could be foiled by ignoring that scientific knowledge is socially constructed and embedded in dominant discourses . While her case deals with decision makers in Brazil, her perspective could also be applied to the US case. It appears that in the face of overwhelming empirical evidence, greater transparency, and even participation, US national decision makers view science as faulty, unfair, subjective and politically imprudent. This leads us to conclude that education, that is to say the increase in knowledge, may not lead to effective policies. On the other hand, Lahsen's chapter does illustrate a potential path. It may be possible to employ science as a structural tool to create a community with shared norms and beliefs.

Finally, once again, William Smith challenges our Western/Northern self-conceptions by categorizing one group of indigenous people as scientists who rely on the cosmos to understand and shape their surroundings through "presence of mind." The contrast between the creators of knowledge from the Totonacs in Chapter 10 and the MIT scientist described by Lahsen in Chapter 8 is startling. These points must be taken into consideration when one makes the assumption that science will provide us the objective knowledge necessary to instigate progressive climate change policies. Increasing knowledge may not be enough, nor might participation or regulation. The question remains, which perspective on science will thrive and will we "attend to distrust related to science and its causes" (Lahsen, p. 189)? The dominant perspective on science that will emerge holds the potential to fundamentally alter our understanding of and responses to climate change.

Social Animals The third opportunity for change emerges in the conceptualization of humans as actors in the climate change drama. When the book is read in its entirety, it appears as if there is a conspiratorial whisper to the means of domination. Who is dominating and what interests underlie their actions?

In some chapters, agents are conceptualized as political or economic beings. For example, it appears that political and power interests dominate the world stage. Several chapters raise the point that the US has played a dominant role in fostering its own political interests by rejecting the Kyoto Protocol and asserting its role as hegemon. This theme is found repeatedly in other sources of climate change literature as well. Many researchers have argued that little will be done to effectively mitigate climate change without the US: "Any global effort to curb anthropogenic global warming will be unsatisfactory without US participation" (Bang, Tjernshaugen and Andresen 2005, 285).

Clearly, the lack of US support for the climate change regime has weakened the international effort to reduce emissions, but is this the only interpretation we should gravitate toward? For example, Cass (2005) argues that once the US abandoned Kyoto, the Europeans quickly shifted their focus and efforts from vetoing emission trading to swiftly adopting this approach in order to keep the protocol alive. While emission trading has its faults and detractors, it has produced a tangible result with the creation of the European Emissions Trading Scheme in 2005. This theme is also repeated in the chapter by Bäckstrand and Lövbrand. In addition, the chapter by Fogel documents aggressive and proactive action on the climate issue within the United States on the substate and NGO levels. While these actions, as she notes, cannot replace US participation in the Kyoto Protocol, it does slightly limit the omnipotent political power of the US national government. Consequently, longing for the return of the US could also be characterized as a myopic view which obscures these important, yet subtle, points.

In contrast, it seems much more productive to ask: How have the US government and key domestic actors defined climate change in such a seemingly radical path in opposition to the rest of the world? Or have they? Perhaps economics is the answer. Several chapters demonstrate similar struggles within states between economic and environmental interests (see the chapters by Cass, Hattori and Pettenger for specific states' policies). Bäckstrand and Lövbrand describe the rise in prominence of the discourse of ecological modernization due to its foundation as a "market-driven" and more "cost-effective" means of addressing climate change. Paterson and Stripple delineate the conceptualization of climate change as an "opportunity" for states (and especially business), who can profit while saving the environment. In this "win-win" world, economics rule as human behavior is driven by getting what we want for the lowest price, with everyone winning. At the same time, Paterson and Stripple show the inherent weaknesses of this perspective, illustrating how economic advantage for some means domination by others. And in the least, such an approach negates the value of nature and perhaps even human life in certain regions of the world. Clearly, knowledge of the economic forces that structure climate change policies is integral to understanding the processes of change. However, is this all, or are other forces being ignored?

In contrast, many chapters offer challenges to political and economic domination. For example, some take up the theme of equity, including discussions of responsibility (the Netherlands push for 50 percent domestic reduction in emissions), environmental stewardship (Japan and nature), and North-South economic and scientific equity (all chapters in the discourse section address this to some degree). Might these challenges to the dominant norms and discourses be lost in limiting our focus to political and economic interests?

One point ties all these issues together. Politics, economics and science gain value and meaning in human interaction, thus in social settings. Ecological Modernization gains power only when people began to change their actions based on its principles. It may exist outside of human agency, as a structure, but it only obtains meaning through agency/humans. Consequently, it will require an alteration of *social* structures and agents' *social* values to bring effective climate change policies, not just political and economic incentives or coercion. Understanding the constructions of climate change requires privileging political, economic *and* social forces.

Participation Finally, the chapters touch on the issue of participation as a source of change. In light of a review of the chapters, we were led to wonder: at what point might dramatic, effective policy be created to confront climate change? What is the tipping point at which effective climate change policy will be made? These questions bring out not only our own subjective interests, but also serve to demonstrate the interplay between power/knowledge and the three principles of constructivism laid out in the Introduction.

For example, Chapter 2 by Cass, demonstrates the different variables that influence norm adoption in three democratic states. As such, the chapter provides an interesting appraisal of a relevant point. Most of the high CO_2 emitting, developed states are democracies. As such, it is assumed that in a democratic system the people have the power to determine the government's actions. This connects directly into the forces of power and knowledge. Who has power in a democracy and what drives their actions? In a democracy, it is customarily assumed that if the people gain knowledge, they then will use this knowledge to influence the government. If we assume this is true, then what is the tipping point at which "the people" will act?

Dr. David Suzuki portrays one interpretation in his summary of Americans' myth-like [our words] understanding of climate change. His study found that most Americans believed that climate change "was a pretty important problem and they were concerned about it," but he was incredulous about the lack of knowledge by the "average Joe" as he stated:

> Apparently … global warming is happening because we've created a hole in the ozone layer, allowing the sun's rays to enter the atmosphere and heat up the earth—or something like that. The cause of the problem is cars, or airplanes, or aerosol cans. No one really knows for sure … People don't get it (Suzuki 2006).

The misunderstanding of climate change that Suzuki describes in the American public consciousness is fundamental to foreshadowing change. The chapters in this volume examine how climate change has been constructed and reconstructed continuously by actors operating at various scales (local, state, international and global). These competing interpretations point to the importance of understanding the construction of knowledge and the role of the public in developed states. Yet, how is new knowledge introduced to the public? What roles do scientists, the media, leaders at all levels, interest groups and NGOs play in constructing knowledge for the public? Perhaps more importantly, what if the people are those that Suzuki describes? How can knowledge become usable (Haas 2004)? Concurrently, can change come from within democracies? Can the people be made to care? This cuts deeply into the ideas of power/knowledge, material/social forces, agent/structure through which change emerges. Is an educated population in a democracy enough? We are left wondering what will tip the scale. Will it come from the people, one leader, a dramatic event, or in slow incremental steps?

In addition, we question whether participation is truly the panacea many assume it to be. Lahsen's chapter suggests that participation may not be enough, if that participation is embedded in a deep distrust of the process. In a warped, cliché sense, just building a "climate change field of dreams and hoping they will come" might not work. In addition, those who may need to participate most based on their vulnerability, the Inuits and the Totonacs for example, are pushed out of the process, their voices muted. Can we imagine a Totonac standing before a legislature in a developed, democratic state asking the leaders to believe in "presence of mind"? In addition, might the participation by a Dutch environmental minister not differ significantly from a Brazilian decision maker?

These questions are particularly germane to the rise of the Civic Environmentalism discourse discussed by Bäckstrand and Lövbrand. One of the principle strengths of the reform version of civic environmentalism is democracy, creating greater spaces for all to participate. However, can this discourse rise to prominence? If so, what form will it take? For some, democracy itself is a social construct (Germino 2001, 793). Thus, we ask, who determines what democracy is and who may participate? Will membership in the dominant discourse be controlled once again by territorially based states, and those with power (developed/democratic states)? Clearly, these many questions cannot be answered at this point, but they serve as suggestions for directions this book may be taken. In sum, participation is highlighted as an upcoming source of change in climate discourses and norms. However, greater research is necessary to understand whether participation is effective to bring change, and what form that participation would need to take.

Summary

Following this brief discussion of the processes of norm and discourse formation and contestation, and the sources of change, we believe it is clear that climate change is more than just a scientifically produced phenomenon, nor is it a political hoax, but rather it is something that is given meaning through the interaction of humans who are touched by its existence. Change, of course, seems to be the fundamental variable in this equation. Thus, we assume that in future climate change negotiations, discourses and norms will continue to be shaped, adopted and rejected, as change is inevitable. Conspicuously, the developed/democratic states have dominated the processes to date, and have had a significant impact on the construction of our understandings of climate change. Their power appears as such in the chapters as a calculated effort to expand their interests and enhance/maintain their identities. Likewise, the marginalized voices have attempted to interpret and shape climate change based on their own interests and identities. Concurrently, gaps appear in the process, allowing for potential change and variance in the process of climate change policy formation.

In conclusion, power (who has it and who does not, and how it is used) and knowledge (what do we know and how we know it) play an integral role in the formation of climate change policies. However, without the tools that constructivism provides, the fluctuations in power and the creation and role of knowledge may be misperceived, misunderstood and ignored. In contrast, recognizing that material and social forces matter, and privileging agents and structures, allows us to gain greater understanding of how (and potentially why) climate change policies have emerged. The theme of change is the consummate end to understanding the social constructivist process, the analytical framework is able to explain (from the normative perspective) or uncover (from the discourse perspective) the processes of social construction, across time and space, and thus signal when, how (and perhaps why) change has taken place. We have strived to demonstrate in this book the paths through which these changes are made. What we offer are suggestions and places to look.

References

Bang, G., A. Tjernshaugen and S. Andresen (2005), Future US Climate Policy: International Re-engagement? *International Studies Perspectives*, 6 (2), pp. 285–303.

Cass, L. (2005), Norm Entrapment and Preference Change: The Evolution of the European Union Position on International Emissions Trading. *Global Environmental Politics*, 5 (2), pp. 38–60.

Checkel, J. (2006), Tracing Causal Mechanisms. *International Studies Review*, 8 (2), pp. 362–370.

Dunn, K. (2006), Examining Historical Representations. *International Studies Review*, 8 (2), pp. 370–381.

Eilperin, J. (2006), Scientists Disagree on Link Between Storms, Warming: Same Data, Different Conclusions. *Washington Post*, (August 21). <http://www.washingtonpoast.com./wp-dyn/content/article/2006/08/19/AR2006081900354_p>.

Germino, D. (2001), Meindert Fennema: Political Theory in Polder Perspective. *The Review of Politics*, 63 (4), pp. 783–804.

Haas, P.M. (2004), When does power listen to truth? A constructivist approach to the policy process. *Journal of European Policy*, 11 (4), pp. 569–592.

Klotz, A. and Lynch, C. (2006), Translating Terminologies. *International Studies Review*, 8 (2), pp. 356–362.

Schlesinger, J. (2005), The Theology of Global Warming. *Wall Street Journal*, (August 8), p. A10.

Suzuki, D. (2006), Public Doesn't Understand Global Warming. *ENN: Environmental News Network*, (16 August). <http:/www.enn.com/comment.html?id=474>.

Index